Elements of
MATHEMATICAL ANALYSIS

R. Agor
Lecturer in Civil Engineering (Retd.)
Technical Education, Delhi

KHANNA BOOK PUBLISHING CO. (P) LTD.
Publisher of Engineering and Computer Books
4C/4344, Ansari Road, Darya Ganj, New Delhi-110002
Phone: 011-23244447-48 **Mobile:** +91-99109 09320
E-mail: contact@khannabooks.com,
sales@khannapublishers.com
Website: www.khannabooks.com, www.khannapublishers.com

ELEMENTS OF MATHEMATICAL ANALYSIS
by R. Agor

ISBN : 978-93-82609-59-9

First Edition : 2015

Published by:
KHANNA BOOK PUBLISHING CO. (P) LTD.
4C/4344, Ansari Road, Daryaganj, New Delhi-110 002
Phone: 011-2324 44 47-48
E-mail: sales@khannapublishers.com
Website: www.khannapublishers.com

Cover Design & Layout by
Akriti Print & Graphics, Delhi

Printed in India by
S.P.S. Printers & Binders, Delhi

Preface to the First Edition

The author is pleased to put forward the first edition of **'Elements of Mathematical Analysis'** to the Students of Engineering and Physics. The object of this book is to set forth in a systematic manner the most important principles, methods and processes used for obtaining numerical results. The book primarily deals with the fundamental principles and processes of the Mathematical Analysis by using numerous solved examples.

Calculations of the numerical solved problems have been gone through by Shri Toshav Agor, B. Tech (Civil), my grand son.

Though, every effort has been made to avoid calculation and printing mistakes, a few of them might have crept due to over right. The author shall be obliged if such errors/omissions are brought to his notice.

Suggestions and criticisms for the improvement of the book shall be gratefully acknowledged for improving the subsequent editions.

The Author expresses his sincere thanks to Shri Mukul Seth, M/s Khanna Book Publishing Co. (P) Ltd., New Delhi 11002.

R. Agor

Brief Table of Contents

Detailed Table of Contents

1 Number System

1.1 INTRODUCTION

There are many systems in which numbers can be expressed. The decimal number system is familiar to us. In this system the base is 10 and the digits are 0, 1, 2, 3, 4, 5, 6, 7, 8, 9. In a number each digit carries a specific weight according to its position. Let us consider a number 575. Here digit 5 in the first position on the right carries a weight of 5 whereas the other 5 on the left in the third position, carries a weight of 500. Each digit has two values, ***i.e.,*** face value and place value.

The following are other popular number systems which are generally used:

1. Binary system	radix	2	2^1
2. Octal system	radix	8	2^3
3. Hexa-decimal system	radix	16	2^4

Most of the digital computers use binary, octal and hexa-decimal systems, with base 2^P where P is 1, 3 and 4 respectively. The digital computers usually use binary number system for reporting an information in its memory because of system's simplicity.

Radix: The total number of digits applicable to any system, is called its **radix.** It is always one more than the highest digit of the system. The digits of various types of number system are as under:

1. Binary system	0, 1
2. Octal system	0, 1, 2, 3, 4, 5, 6, 7
3. Decimal system	0, 1, 2, 3, 4, 5, 6, 7, 8, 9
4. Hexa-decimal system	0, 1, 2, 3, 4, 5, 6, 7, 8, 9, A, B. C, D, E, F.

Illustration: The numbers in various systems are expressed as unde:

1. Binary system	1010011
2. Octal system	12461
3. Decimal system	3579
4. Hexadecimal system	5B73

By using the digits 0, 1, 2, 3, 4, 5, 6, 7, the numbers in octal system are as under:

0, 1, 2, 3, 4, 5, 6, 7, [8, 9,] 10, 11, 12, 13, 14, 15, 16, 17, [18, 19,] 20, 21, 22, 23, 24, 25, 26, 27, [28, 29,] 30, 31, 32, 33, 34, 35, 36, 37, [38, 39,] 40, 41, 42, 43, 44, 45, 46, 47, [48, 49,] 50, 51, 52, 53, 54, 55, 56, 57, [58, 59,] 60, 61, 62, 63, 64, 65, 66, 67, [68, 69,] 70, 71, 72, 73, 74, 75, 76, 77, [78, 79, 80, 81, 82, 83, 84, 85. 86, 87, 88, 89, 90, 91, 92, 93, 94, 95, 96, 97, 98, 99,] 100. It may be noted that in octal system, the numbers consist of the digits 0 to 7.

1.2 CONVERSION OF A NUMBER FROM ANY OTHER SYSTEM TO DECIMAL SYSTEM

Any number in any system may be expressed by the following general expression:

$$N = A_m A_{m-1} A_{m-2} A_{m-3} \dots\dots\dots A_2 A_1 A_0$$

or

$$N = A_m R^m + A_{m-1} \cdot R_{m-1} + A_{m-2} R^{m-2} + \dots\dots\dots + A_2 E^2 + A_1 R^1 + A_0 R^0$$

where R is the radix and m is the place position.

Let us study the formation of the number 1970 in decimal system.

$$\begin{aligned}(1970)_{10} &= 1 \times 10^3 + 9 \times 10^2 + 7 \times 10^1 + 0 \times 10^0 \\ &= 1000 + 900 + 70 + 0 \\ &= 1970\end{aligned}$$

(*i*) Convert binary number $(111)_2$ into decimal system.

$$\begin{aligned}(111)_2 &= 1 \times 2^2 + 1 \times 2^1 + 1 \times 2^0 \\ &= 4 + 2 + 1 \\ &= (7)_{10}. \quad \textbf{Ans.}\end{aligned}$$

(*ii*) Convert octal number $(1654)_8$ into decimal system

$$\begin{aligned}(1654)_8 &= 1 \times 8^3 + 6 \times 8^2 + 5 \times 8^1 + 4 \times 8^0 \\ &= 512 + 384 + 40 + 4 = (940)_{10}. \quad \textbf{Ans.}\end{aligned}$$

(*iii*) Convert hexa-decimal number $(123)_{16}$ to decimal system

$$\begin{aligned}(123)_{16} &= 1 \times 16^2 + 2 \times 16^1 + 3 \times 16^0 \\ &= 256 + 32 + 3 = (291)_{10}. \quad \textbf{Ans.}\end{aligned}$$

1.3 CONVERSION OF A NUMBER FROM DECIMAL SYSTEM TO OTHER SYSTEMS

The number of any system, may be converted to decimal system by the following two methods:

1. Remainder method
2. Power method

1. **Remainder method.** The conversion of a number from decimal system to other system may be done as under:

(*i*) Divide the given number of decimal system by the radix R of the desired system. We get a quotient Q_1 and a remainder R_1.

(*ii*) Divide the quotient Q_1 by the radix R again to get quotient Q_2 and a remainder R_2.

(*iii*) Divide the quotient Q_2 by the radix R again to get quotient Q_3 and a remainder R_3.

....

....

....

(*n*) Divide the quotient Q_{n-1} by the radix R to get the quotient Q_n and a remainder R_n.

The process of division of the successive quotients by the radix R of the desired number system is repeated till the quotient becomes less than the radix R. The required number is obtained by writing the last quotient first and then remainders in the reverse order, ***i.e.,***

$$Q_n \cdot R_n \cdot R_{n-1} \cdot R_{n-2} \text{ .... } R_3 \cdot R_2 \cdot R_1$$

The method of conversion of the numbers from decimal system to other systems is explained in the following solved examples.

Example. 1.1. *Convert $(940)_{10}$ to binary system.*

Solution: The radix of the binary system is 2.

		Remainder
2	940	
2	470	0
2	235	0
2	117	1
2	58	1
2	29	0
2	14	1
2	7	0
2	3	1
	1	1

i.e., $(940)_{10} = (1110101100)_2$. **Ans.**

Example 1.2. *Convert $(2040)_{10}$ to octal system.*

Solution: The radix of the octal system is 8.

Divisor	Number	Remainder
8	2040	
8	255	0
8	31	7
	3	7

i.e., $(2040)_{10} = (3770)_8$. **Ans.**

Example 1.3. *Convert $(2040)_{10}$ to hexa-decimal system.*

Solution: The radix of the hexa-decimal system is 16.

Divisor	Number	Remainder
16	2040	
16	127	8
	7	15

i.e., $(2040)_{16} = (7E8)_{16}$ **Ans.**

where E represents 15 in the hexa-decimal system.

Example 1.4. *Obtain the octal equivalent of $(3964)_{10}$.*

Solution: Radix ol the octal system is 8.

Divisor	Number	Remainder
8	3964	
8	495	4
8	61	7
	7	5

$\therefore$ $(3964)_{10} = (7574)_8$. **Ans.**

2. **Power method.** Conversion of a number from the decimal system to other system, may be done as under:

(*i*) Subtract the highest number which is obtained by raising the radix to the power of the desired system. Let the remainder be R_1.

(*ii*) Subtract the next highest number from the remainder R_1.

(*iii*) Repeat the procedure till the remainder is zero.

(*iv*) Write the multiplication factors in the sequential order from top to bottom.

The method of conversion of the numbers from decimal system to other systems by power method is explained in the following solved examples.

Example 1.5. *Convert $(300)_{10}$ to the binary system.*

Solution: Here radix of binary system is 2.

2^8	2^7	2^6	2^5	2^4	2^3	2^2	2^1	2^0
256	128	64	32	16	8	4	2	1

We note that 256 is the highest number which is less than 300, the given number.

The procedure of subtraction is done as under:

$$\begin{array}{rll} 300 & & \\ -256 & 2^8 & = 1 \times 2^8 \\ \hline 44 & & \\ -32 & 2^5 & = 1 \times 2^5 \\ \hline 12 & & \\ -8 & 2^3 & = 1 \times 2^3 \\ \hline 4 & & \\ -4 & 2^2 & = 1 \times 2^2 \\ \hline 0 & & \end{array}$$

As powers of $2^7, 2^6, 2^4, 2^1, 2^0$ are not represented, these values may be placed in their respective places after multiplying with zero, *i.e.*,

$1 \times 2^8, 0 \times 2^7, 0 \times 2^6, 1 \times 2^5, 0 \times 2^4, 1 \times 2^3, 1 \times 2^2, 0 \times 2^1, 0 \times 2^0$

i.e., $(300)_{10} = (100101100)_2$. **Ans.**

Example 1.6. *Convert $(300)_{10}$ to octal system.*

Solution: The radix of the number is 8

.........8^5	8^4	8^3	8^2	8^1	8^0
.........32768	4096	512	64	8	1

We note that 256 is the highest number i.e. 4×8^2.

$$\begin{array}{rl} 300 & \\ -256 & 4 \times 8^2 \\ \hline 44 & \\ -40 & 5 \times 8^1 \\ \hline 4 & \\ -4 & 4 \times 8^0 \\ \hline \end{array}$$

i.e., $(300)_{10} = (454)_8$. **Ans.**

Example 1.7. *Convert $(1654)_8$ into binary system.*

Solution: We will first convert the given number to decimal system and then will convert the number from decimal system to binary system.

i.e., $(1654)_8 = 1 \times 8^3 + 6 \times 8^2 + 5 \times 8^1 + 4 \times 8^0$

$$= 512 + 384 + 40 + 4$$
$$= (940)_{10}$$

Now, convert $(940)_{10}$ to the binary system as explained in example 1.1.

i.e., $(1654)_8 = (940)_{10} = (1110101100)_2$. **Ans.**

1.4 CONVERSION OF A FRACTION OF DECIMAL SYSTEM TO OTHER SYSTEMS

The fraction part of a decimal system may be converted to other systems as under:

(*i*) Multiply the fraction of the decimal system by the radix R of the required system.

(*ii*) Let I_1 be the resulting integer and f_1 be the fraction.

(*iii*) Multiply f_1 by the radix again.

(*iv*) Let I_2 be the integer and f_2 be the fraction.

(*v*) Repeat the above steps till the fraction part vanishes.

(*vi*) The required fraction part in the required system with radix R is $I_1, I_2, I_3, I_4 \ldots\ldots\ldots I_n$.

The procedure of converting the fractions of decimal system to other system, is explained in the following solved examples.

Example 1.8. *Obtain the binary equivalent of* $(9.375)_{10}$.

Solution: Integer number is 9

Divisor	Number	Remainder
2	9	
2	4	1
2	2	0
	1	0

i.e., $9_{(10)} = (1001)_2$

Fraction part is $0.375_{(10)}$ — Integer

	Integer
$.375 \times 2 = 0.750$	0
$.75 \times 2 = 1.50$	1
$.5 \times 2 = 1.0$	1

$\therefore$ Fraction part $= (0.011)_2$

$\therefore$ Required number $= (1001.011)_2$. **Ans.**

Example 1.9. *Convert* $(17.35)_{10}$ *to binary form.*

Solution: Integer number is 17.

Divisor	Dividend	Remainder
2	17	
2	8	1
2	4	0
2	2	0
	1	0

$(17)_{10} = (10001)_2$

Fraction part is. 35. Integer

	Integer
$0.35 \times 2 = 0.70$	0
$0.7 \times 2 = 1.4$	1
$0.4 \times 2 = .8$	0
$0.8 \times 2 = 1.6$	1
$0.6 \times 2 = 1.2$	1
$0.2 \times 2 = 0.4$	0

$\therefore \quad (17.35)_{10} = (10001.010110...)_2$. **Ans.**

Example 1.10. *Convert $(0.25)_{10}$ to the binary system.*

Solution: $0.25 \times 2 = 0.50 \qquad I_0 = 0$

$0.50 \times 2 = 1.00 \qquad I_2 = 1$

$\therefore \quad (0.25)_{10} = (0.01)_2$. **Ans.**

Example 1.11. *Convert $(0.2)_{10}$ to binary form.*

Solution: Radix = 2 Integer

	Integer
$0.2 \times 2 = 0.4$	0
$0.4 \times 2 - 0.8$	0
$0.8 \times 2 = 1.6$	1
$0.6 \times 2 = 1.2$	1
$0.2 \times 2 = 0.4$	0

$\therefore \quad (0.2)_{10} = (0.00110...)_2$. **Ans.**

Example 1.12. *Convert $(634.640625)_{10}$ to octal equivalent.*

Solution: (*i*) Integer part is $(634)_{10}$

Divisor	Dividend	Remainder
8	634	
8	79	2
8	9	7
	1	1

$(634)_{10} = (1172)_8$.

(*ii*) Fraction part is .640625 Integer

	Integer
$0.640625 \times 8 = 5.125$	5
$0.125 \times 8 = 1.0$	1

$\therefore$ Octal equivalent $= (1172.51)_8$. **Ans.**

Example 1.13. *Convert $(0.0625)_{10}$ to the octal system.*

Solution: Radix is 8.

$$0.0625 \times 8 = 0.5 \qquad I_1 = 0$$
$$0.5 \times 8 = 4.0 \qquad I_2 = 4$$
$$\therefore \quad (0.0625)_{10} = (0.04)_8. \quad \textbf{Ans.}$$

1.5 CONVERSION OF A FRACTION OF ANY SYSTEM WITH RADIX R TO THE DECIMAL SYSTEM

The following steps are taken for converting a fraction of any system to the decimal system.

(*i*) Multiply the digits after the point (i.e. binary point, octal point or hexa-decimal point) by R^{-1}, R^{-2}, R^{-3}, ..., etc. respectively where R is the radix of the given system.

(*ii*) Simplify the fractions thus obtained and add them together to get the required fraction of the decimal system.

Example 1.14. *Convert $(0.01)_2$ to the decimal system.*

Solution: Radix is 2.

$$(0.01)_2 = 0 \times 2^{-1} + 1 \times 2^{-2}$$
$$= 0 + 0.25$$
$$\text{or} \quad (0.01)_2 = (0.25)_{10}. \quad \textbf{Ans.}$$

Example 1.15. *Convert $(0.135)_8$ to the decimal system.*

Solution. Radix is 8.

$$(0.135)_8 = 1 \times 8^{-1} + 3 \times 8^{-2} + 5 \times 8^{-3}$$
$$= \frac{1}{8} + \frac{3}{64} + \frac{5}{512}$$
$$= \frac{93}{512} = (0.1816406)_{10}. \quad \textbf{Ans.}$$

Example 1.16. *Convert $(17.35)_{10}$ into binary system.*

Solution: Integer number is 17.

2	17	Remainder
2	8	1
2	4	0
2	2	0
	1	0

$$(17)_{10} = (10001)_2$$

Fraction part is. 35. Integer

	Integer
$.35 \times 2 = 0.70$	0
$.7 \times 2 = 1.4$	1
$.4 \times 2 = .8$	0
$.8 \times 2 = 1.6$	1
$.6 \times 2 = 1.2$	1
$.2 \times 2 = 0.4$	0

$\therefore \quad (17.35)_{10} = (10001.010110...)_2.$ **Ans.**

Example 1.17. *Convert $(0.25)_{10}$ to the binary system.*

Solution:

$$0.25 \times 2 = 0.50 \qquad I_0 = 0$$

$$0.50 \times 2 = 1.00 \qquad I_2 = 1$$

$\therefore \quad (0.25)_{10} = (0.01)_2.$ **Ans.**

Example 1.18. *Convert $(0.2)_{10}$ to binary form.*

Solution: Radix = 2

	Integer
$0.2 \times 2 = 0.4$	0
$0.4 \times 2 - 0.8$	0
$0.8 \times 2 = 1.6$	1
$0.6 \times 2 = 1.2$	1
$0.2 \times 2 = 0.4$	0

$\therefore \quad 0.2_{10} = (0.00110...)_2.$ **Ans.**

Example 1.19. *Convert the following decimal number to hexadecimal system.*

735

Solution:

16	735	15	Now 15 = F in hexadecimal
16	45	13	13 = D in hexadecimal
	2		2 = 2 in hexadecimal

$\therefore \quad (735)_{10} = (2DF)_{16}$ **Ans.**

Example 1.20. *Convert the decimal number 41.6875 to a binary system.*

Solution:

Integer Part 41

2	41	1
2	20	0
2	10	0
2	5	0
2	2	1
	1	0

Fractional Part 0.6875

$0.6875 \times 2 = 1.3750$	1
$0.3750 \times 2 = 0.7500$	0
$0.7500 \times 2 = 1.5000$	1
$0.5000 \times 2 = 1.0000$	1

$(0.6875)_{10} = (1011)_2$

$$(41)_{10} = (101001)_2$$

$\therefore$ $(41.6875)_{10} = (101001.1011)2$ **Ans.**

Example 1.21. *Convert the decimal number 153.513 into an octal number.*

Solution: Integer part is obtained by dividing 153 by 8 remainders

8	153	
8	19	1
	2	3

$\therefore$ $(153)_{10} = (231)_8$

The fraction part is obtained by multiplying the fraction part successively by 8

$$0.513 \times 8 = 4.104$$
$$0.104 \times 8 = 0.832$$
$$0.832 \times 8 = 6.656$$
$$0.656 \times 8 = 5.248$$
$$0.248 \times 8 = 1.984$$

$\therefore$ $(153.513)_{10} = (231.40651)_8$ **Ans.**

Example 1.22. *Convert the following numbers as instructed:*

$$(375.25)_{10} = (?)_8$$

Solution: $(375.25)_{10}$ may be converted to octal as under:

Integer

8	375	
8	46	7
	5	6

Fraction

$0.25 \times 8 = 2.002$

$(0.25)_{10} = (0.2)_8$

$\therefore$ $(375)_{10} = (567)_8$

or $(375.25)_{10} = (567.2)_8$ **Ans.**

Example 1.23. *Convert $(3.75)_8$ to binary system.*

Solution: The integer part 3 = $(11)_2$.

The fraction part $(0.75)_8$ may first be converted to the decimal system.

i.e., $$(0.75)_8 = 7 \times 8^{-1} + 5 \times 8^{-2}$$

$$= \frac{7}{8} + \frac{5}{4} = \frac{61}{64} = (0.953125)_{10}$$

$$\therefore \quad (0.75)_8 = (0.953125)_{10}.$$

Now convert $(0.953125)_{10}$ to the binary system as under:

$$\begin{aligned} 0.953125 \times 2 &= 1.906250 & I_1 &= 1 \\ 0.906250 \times 2 &= 1.18125 & I_2 &= 1 \\ 0.8125 \times 2 &= 1.6250 & I_3 &= 1 \\ 0.625 \times 2 &= 1.250 & I_4 &= 1 \\ 0.25 \times 2 &= 0.50 & I_5 &= 0 \\ 0.50 \times 2 &= 1.00 & I_6 &= 1 \end{aligned}$$

$$\therefore \quad (0.953125)_{10} = (0.111101)_2$$

$$\therefore \quad (3.75)_8 = (11.111101)_2. \quad \textbf{Ans.}$$

1.6 DIRECT METHOD OF CONVERSION A NUMBER FROM ONE SYSTEM TO THE OTHER

Whenever the radix of one numbering system can be expressed in powers of the radix of the other system, the system can be converted directly by forming pairs of 3 or 4 as the case might be, by noting the power of the. radix. For converting the digits of decimal system into binary system refer to Table 1.1.

Table 1.1

Decimal digit	*Binary equivalent*
0	0000
1	0001
2	0010
3	0011
4	0100
5	0101
6	0110
7	0111
8	1000
9	1001

From the above table, it is evident that a decimal digit may have at the most lour bits.

For converting the digits of binary system into hexa-decimal system refer to Table 1.2.

Table 1.2

Decimal	*Binary*	*Hexa-decimal*	*Decimal*	*Binary*	*Hexa-decimal*
0	0000	0	8	1000	8
1	0001	1	9	1001	9
2	0010	2	10	1010	A
3	0011	3	11	1011	B
4	0100	4	12	1100	C
5	0101	5	13	1101	D
6	0110	6	14	1110	E
7	0111	7	15	1111	F

For converting the digits of binary system to octal system, refer to Table 1.3.

Table 1.3

Octal	*Binary*	*Octal*	*Binary*
0	000	4	100
1	001	5	101
2	010	6	110
3	011	7	111

Example 1.24. *Convert $(1352)_8$ to binary system.*

Solution: Here radix $8 = 2^3$

Form pairs of each digit of the octal system with binary system (Table 1.3).

$$\therefore \quad (1352)_8 = (001, 011, 101, 010)_2. \quad \textbf{Ans.}$$

Example 1.25. *Convert $(CD578)_{16}$ into binary system.*

Solution: Here $16 = 2^4$; power is 4

Form pairs of each digit of the hexa-decimal system into binary equivalents.

$$\begin{aligned} C &= 12 & (12)_2 &= 1100 \\ D &= 13 & (13)_2 &= 1101 \\ & & (5)_2 &= 0101 \\ & & (7)_2 &= 0111 \\ & & (8)_2 &= 1000 \end{aligned}$$

$\therefore$ The required number is $(1100, 1101, 0101, 0111, 1000)_2$. **Ans.**

Example 1.26. *Convert the binary number 1010011001.1011001 to octal system.*

Solution: Radix 8 = 2^3

001	010	011	001	.	101	100	100
1	2	3	1	.	5	4	4

$\therefore$ $(001010011001.101100100)_2 = (1231.544)_8$. **Ans.**

Example 1.27. *Convert $(0.625)_{10}$ to binary system.*

Solution: Radix is 2.

$0.625 \times 2 = 1.250 \quad I_1 = 1$

$0.250 \times 2 = 0.500 \quad I_2 = 0$

$0.500 \times 2 = 1.000 \quad I_3 = 1$

$\therefore \quad (0.625)_{10} = (.101)2$. **Ans.**

Example 1.28. *Convert the following numbers from one system to other as indicated against each.*

(*i*) *$(7312)_8$ to decimal system.*
(*ii*) *$(101101)_2$ to decimal system.*
(*iii*) *$(689B)_{16}$ to decimal system.*
(*iv*) *$(3271)_8$ to binary system.*
(*v*) *$(123)_{16}$ to binary system.*
(*vi*) *$(110101101)_2$ to octal system.*
(*vii*) *$(7715)_8$ to binary system.*
(*viii*) *$(1101.1111)_2$ to hexa-decimal system.*
(*ix*) *$(274655)_8$ to binary system.*
(*x*) *$(0.735)_{16}$ to binary system.*

Solution:

(*i*) $(7312)_8 = 7 \times 8^3 + 3 \times 8^2 + 1 \times 8^1 + 2 \times 8^0$
$= 7 \times 512 + 3 \times 64 + 1 \times 8 + 2$
$= 3584 + 192 + 8 + 2$
$= (3786)_{10}$. **Ans.**

(*ii*) $(101101)_2 = 1 \times 2^5 + 0 \times 2^4 + 1 \times 2^3 + 1 \times 22 + 0 \times 21 + 1 \times 2^0$
$32 + 0 + 8 + 4 + 0+1$
$= (45)_{10}$. **Ans.**

(*iii*) $(689B)_{16} = 6 \times 16^3 + 8 \times 16^2 + 9 \times 16^1 + 11 \times 16^0$ ($\because$ B = 11)
$= 6 \times 4096 + 8 \times 256 + 9 \times 16 + 11$
$= 24576 + 2048 + 144 + 11$
$= (26779)_{10}$. Ans.

(*iv*) $(3271)_8 = 3 \times 8^3 + 2 \times 82 + 7 \times 81 + 1 \times 8^0$
$= 1536 + 128 + 56+1$
$= (1721)_{10}$

Now (1721)10 may be converted into binary system.

2	1721	Remainder
2	860	1
2	430	0
2	215	0
2	107	1
2	53	1
2	26	1
2	13	1
2	6	1
2	3	0
	1	1

$\therefore \quad (3271)_8 = (1721)_{10} = (11010111001)_2$. **Ans.**

(*v*) $(123)_{16}$ to binary system

$$= 1 \times 162 + 2 \times 16^1 + 3 \times 16^0$$

$$= 256 + 32 + 3 = (291)_{10}$$

Now, convert $(291)_{10}$ to the binary system.

2	291	Remainder
2	145	1
2	72	0
2	36	0
2	18	0
2	9	0
2	4	1
2	2	0
	1	0

$\therefore \quad (123_{16} = (291)_{10} = (1\ 0010\ 0011)_2$. **Ans.**

(*vi*) $(110101101)_2$ to octal system.

First convert the given binary number into decimal system *i.e.*, $(110101101)_2$

$$= 1 \times 2^8 + 1 \times 2^7 + 0 \times 2^6 + 1 \times 2^5 + 0 \times 2^4 + 1 \times 2^3 + 1 + 0 \times 21 + 1 \times 2^0$$

$$= 256 + 128 + 0 + 32 + 0 + 8 + 4 + 0 + 1$$

$$= (429)_{10}$$

Now convert $(429)_{10}$ to octal system

8	429	Remainder
8	53	5
	6	5

$\therefore \quad (110101101)_2 = (429)_{10} = (655)_8$. **Ans.**

(*vii*) $(7715)_8$ to binary system.

First convert the given octal number into decimal system.

$$(7715)_8 = 7 \times 8^3 + 7 \times 8^2 + 1 \times 81 + 5 \times 8^0$$
$$= 7 \times 512 + 7 \times 64 + 1 \times 8 + 5$$
$$= 3584 + 448 + 8 + 5$$
$$= (4045)_{10}$$

Now, convert $(4045)_{10}$ into binary system.

2	4045	Remainder
2	2022	1
2	1011	0
2	505	1
2	252	1
2	126	0
2	63	0
2	31	1
2	15	1
2	7	1
2	3	1
	1	1

$\therefore \quad (7715)_8 = (4045)_{10} = (111111001101)_2$. **Ans.**

(*viii*) $(1101.1111)_2$ to hexa-decimal system

(*a*) Integer part is 1101

$$= 1 \times 2^3 + 1 \times 2^2 + 0 \times 2^1 + 1 \times 2^0$$
$$= 8 + 4 + 0 + 1$$
$$= 13$$

$\therefore \quad (1101)_2 = (13)_{10} = (D)_{16} \qquad (\because \ D = 13)$

Fraction part is .1111

$$= 1 \times 2^{-1} + 1 \times 2^{-2} + 1 \times 2^{-3} + 1 \times 2^{-4}$$
$$= \frac{1}{2} + \frac{1}{4} + \frac{1}{8} + \frac{1}{16}$$

$$= \frac{15}{16} \times = (0.9375)_{10}$$

$$\therefore \quad (1101.1111)_2 = (13.9375)_{10}$$

$$= (D.F)_{16}. \quad \textbf{Ans.}$$

(*ix*) (274655)8 to binary system.

First convert the given number into decimal system. *i.e.*, $(274655)_8$

$$= 2 \times 8^5 + 7 \times 8^4 + 4 \times 8^3 + 6 \times 8^2 + 5 \times 8^1 \times 5 \times 8^0$$

$$= 65536 + 28672 + 2048 + 384 + 40 + 5$$

$$= (96685)_{10}$$

Now, convert $(96685)_{10}$ to binary system.

2	96685	Remainder
2	48342	1
2	24171	0
2	12085	1
2	6042	1
2	3021	0
2	1510	1
2	755	0
2	377	1
2	188	1
2	94	0
2	47	0
2	23	1
2	11	1
2	5	1
2	2	1
	1	0

$$\therefore \quad (274655)_8 = (96685)_{10} = (10111100110101101)_2. \quad \textbf{Ans.}$$

(*x*) $(0375)_{16}$ to binary system.

First convert the given fraction into decimal fraction.

$$(0375)_{16} = 3 \times 16^{-1} + 7 \times 16^{-2} + 5 \times 16^{-3}$$

$$= \frac{3}{16} + \frac{7}{256} + \frac{5}{4096}$$

$$= \frac{768 + 112 + 5}{4096} = \frac{885}{4096} = (0.2160644)_{10}$$

Now convert (0.2160644).() into binary system Integer

		Integer
0.2160644 × 2	= .4321288	0
0.4321288 × 2	= .8642576	0
0.8642576 × 2	= 1.7285152	1
0.7285152 × 2	= 1.4570304	1
0.45703125 × 2	= 0.9140608	0
0.9140608 × 2	= 1.8281216	1
0.8281216 × 2	= 1.6562432	1
0.6562432 × 2	= 1.3124864	1
0.3124864 × 2	= 0.6249728	0
0.6249728 × 2	= 1.250	1
0.250 × 2	= 0.500	0
0.50 × 2	= 1.000	1

$\therefore \quad (0.375)_{16} = (0.2160644)_{10}$

$= (00110111\ 0101\ )_2$. **Ans.**

1.7 BINARY CALCULATIONS

The addition, subtraction, multiplication and division of binary numbers, may be done as in the case of decimal system.

(1) **Addition.** While adding the binary numbers, the following rules are followed:

0 + 0 = 0

0 + 1 = 1

1 + 0 = 1

1 + 1 = 0 with 1 carry over.

Example 1.29. *Add* $(110101)_2$ *and* $(101111)_2$.

Solution:

$$\begin{array}{r} 110101 = 53 \\ +\ 101111 = 47 \\ \hline 1100100 = 100. \end{array}$$ **Ans.**

Example 1.30. *Add* $(110101101)_2$, $(111011101)_2$, $(10010U01)_2$, $(111110111)_2$ *and* $(111111001101)_2$.

Solution:

	Decimal equivalent
1 1 0 1 0 1 1 0 1	4 2 9 1
1 1 1 0 1 1 1 0 1	4 7 7
0 0 0 1 1 1 1 1 1	6 3
1 0 0 1 0 1 1 0 1	3 0 1
1 1 1 1 1 0 1 1 1	5 0 3
1 1 1 1 1 1 0 0 1 1 0 1	4 0 4 5
1 0 1 1 0 1 0 1 1 1 0 1 0	$(5818)_{10}$ **Ans.**

Example 1.31. *Perform the following operation. 26 + 37 using BCD number system.*

Solution: We shall first convert 26 and 37 to binary number system and then we will add.

2	26	Remainder
2	13	0
2	6	1
2	3	0
	1	1

2	37	Remainder
2	18	1
2	9	0
2	4	1
2	2	0
	1	0

i.e., $(26)_{10} = (11010)_2$ *i.e.,* $(37)_{10}$ = 100101

$$\begin{array}{rcr} 26 & = & 11010 \\ +\ 37 & = & 100101 \\ \hline 63 & & 111111 \end{array}$$

$$\begin{aligned}(111111)_2 &= 1 \times 2^5 + 1 \times 2^4 + 1 \times 2^3 + 1 \times 2^2 + 1 \times 2^1 + \times 2^0 \\ &= 32 + 16 + 8 + 4 + 2 + 1 \\ &= (63)_{10} \quad \textbf{O.K.}\end{aligned}$$

(2) **Subtraction.** While subtracting the binary numbers, the following rules may be followed:

$$\begin{aligned} 0 - 0 &= 0 \\ 1 - 0 &= 1 \\ 1 - 1 &= 0 \\ 0 - 1 &= 1 \text{ with one borrow.}\end{aligned}$$

Example 1.32. *Subtract* $(100101100)_2$ *from* $(1110101010)_2$.

Solution.

```
  1 1 1 0 1 0 1 0 1 0
    1 0 0 1 0 1 1 0 0
  -------------------
  1 0 0 1 1 1 1 1 1 0.      Ans.
```

(3) **Multiplication.** While multiplying the binary numbers, the following rules may be followed:

$$0 \times 0 = 0$$
$$0 \times 1 = 0$$
$$1 \times 1 = 1.$$

Example 1.33. *Multiply* $(111)_2$ *by* $(101)_2$.

Solution:

```
      111                 7
    × 101               × 5
    -----              ----
      111                35
     000
    111
   ------
   100011
```

$\therefore \quad (111)_2 \times (101)_2 = (100011)_2$. **Ans.**

Check: $(100011)_2 = 1 \times 2^5 + 0 \times 2^4 + 0 \times 2^3 + 0 \times 2^2 + 1 \times 2^1 + 1 \times 2^0$

$= 32 + 0 + 0 + 0 + 2 + 1$

$= 35$. **O.K.**

(4) **Division.** While dividing a binary number by another binary number, the rules applicable to the decimal system may be followed and also obeying the rules of addition, subtraction and multiplication of the binary system.

Example 1.34. *Divide* $(100011)_2$ *by* $(101)_2$.

Solution:

```
101) 100011 (111
     101
     ------
      111
      101
     ------
      101
      101
     ------
        0
```

i.e., $(100011)_2 \div (101)_2 = (111)_2$. **Ans.**

Example 1.35. *Multiply $47_{(8)}$ by $52_{(8)}$ and give the result as octal number.*

Solution: We shall convert $47_{(8)}$ and $52_{(8)}$ into decimal system, multiply the numbers in decimal system and, convert the product obtained in decimal system to octal system.

$$47_{(8)} = 4 \times 8^1 + 7 \times 8^0 = 39_{(10)}$$
$$2_{(8)} = 5 \times 81 + 2 \times 8^0 = 42_{(10)}$$
$$\therefore \quad 39 \times 42 = (1638)_{10}$$

8	1638	Remainder
8	204	6
8	25	4
	3	1

$$\therefore \quad 47_{(8)} \times 52_{(8)} = (3146)_8. \quad \textbf{Ans.}$$

Direct Method

```
  5 2
 ----
  4 7
4 4 6
2 5 0
------
3 1 4 6
```

$$\therefore \quad 52_{(8)} \times 47_{(8)} = (3146)_8. \quad \textbf{Ans.}$$

Example 1.36. *Explain how a binary number 1001 is multiplied by another binary number 1101 in the ALU unit of digital computer. Also, illustrate the division of 1111 by 11 in binary form. Show the schematic circuits.*

Solution: (*a*) **Multiplication**

$$1001 = 1 \times 2^3 + 0 \times 2^2 + 0 \times 2^1 + 1 \times 2^0 = 9$$
$$1101 = 1 \times 2^3 + 1 \times 2^2 + 0 \times 2^1 + 1 \times 2^0 = 13$$
i.e., $$9 \times 13 = 117$$

```
       1001
     × 1101
    -------
       1001
      0000
     1001
    1001
    -------
    1110101   Ans.
```

Check: $1110101 = 1 \times 2^6 + 1 \cdot \times 2^5 + 1 \times 2^4 + 0 \times 2^3 + 1 \times 2^2 + 0 \times 0^1 + 1 \times 2^0$

$= 64 + 32 + 16 + 0 + 4 + 0 + 1 = 117.$ **O.K.**

(4) Division

$$\begin{array}{r} 101 \\ 11)\overline{1111} \\ 11 \\ \hline 11 \\ 11 \\ \hline 0 \end{array}$$

Check: $1111 = 1 \times 2^3 + 1 \times 2^2 + 1 \times 2^1 + 1 \times 2^0 = 15$

$11 = 1 \times 21 + 1 \times 2^0 = 3$

i.e., $15 + 3 = 5 = (101)_2$. **Ans.**

1.8 MULTIPLICATION OF TWO DIGITS HAVING ANY INDEX

The multiplication of two digits having any index, may be done as under:

Multiply the two digits in decimal and if the product is more than the radix, take a carry equal to quotient diving the product by the radix. Use the remainder as the first or least significant position.

Example 1.37. *Multiply* $(47)_8 \times (25)_8$.

Solution:

Octal Product	*Equialent Decimal Product*
47	39
25	21
303	39
116	78
$(1463)_8$	819

$= 1 \times 8^3 + 4 \times 8^2 + 6 \times 8^1 + 3 \times 8^0$

$= 512 + 256 + 48 + 3 = 819$ **Ans.**

Example 138. *Perform the following multiplication of octal numbers*

(*i*) *153 × 35*

(*ii*) *15.2 × 5.4*

Solution:

(*i*)

$$\begin{array}{r} 153 \\ 35 \\ \hline 1027 \\ 501 \\ \hline 6037 \\ \hline \end{array}$$ **Ans.**

(*ii*)

$$\begin{array}{r} 15.2 \\ 5.4 \\ \hline 650 \\ 1022 \\ \hline 110.70 \\ \hline \end{array} \quad \text{in octal system}$$

Check:

$$(15.2)_8 = 1 \times 8^1 + 5 \times 8^0 \,.\, 2 \times 8^{-1}$$

$$= 8 + 5 \,.\, \frac{1}{4}$$

$$= 13.25$$

$$(5.4)_8 = 5 \times 8^\circ \,.\, 4 \times 8^{-1}$$

$$= 5 \times \frac{4}{8}$$

$$= 5.5$$

$$\therefore \quad 13.25 \times 5.5 = \begin{array}{r} 13.25 \\ 5.5 \\ 6625 \\ \hline 6625 \\ \hline 72.875 \\ \hline \end{array}$$

$$(110.7)_8 = 1 \times 8^2 + 1 \times 8^1 + 0 \times 8^0 \,.\, 7 \times 8^{-1}$$

$$= 64 + 8 + 0 \,.\, \frac{7}{8}$$

$$= 72.875 \quad \textbf{Ans.}$$

Example 1.39. *Perform the following multiplications of hexadecimal numbers*

(i) 15 × B *(ii) A 12 × 6*

Solution:

(*i*) 15 × B

In hexadecimal number system B = 11

$$\therefore \quad \begin{array}{r} 15 \\ 11 \\ \hline \textcircled{14}\ 7 \end{array}$$

But $14 = E$

$\therefore$ $(15 \times B)_{16} = (E7)_{16}$

Check: $15 = 1 \times 16^1 + 5 \times 16^0 = 21_{10}$

$B = 11_{10}$

$\therefore$ $21 \times 11 = (231)_{10}$

But $E7 = 14 \times 16^1 + 7 \times 10^0 = 224 + 7 = (231)_{10}$ **Ans.**

(*ii*) A 12 × 6

$$\begin{array}{r} 1012 \\ 6 \\ \hline 3.12.6.12 \end{array}$$

∴ A = 10 in hexadecimal system

or $(3.C.6.C)_{16} = 3 \times 16^3 + 12 \times 16^2 + 6 \times 16^1 + 12 \times 16^0$

$= 12288 + 3072 + 96 + 12 = (15468)_{10}$

Check: $(A12)_{16} = 10 \times 16^2 + 1 \times 16^1 + 2 \times 16^0 = 2560 + 16 + 2$

$= (2578)_{10}$

$(6)_{16} = (6)_{10}$

∴ $2578 \times 6 = (15468)_{10}$ **Ans.**

Example 1.40. *Convert the following decimal numbers to binary:*
(i) $(0.3)_{10}$ *and (ii)* $(0.6)_{10}$

Solution:

	Integer
$0.3 \times 2 = 0.6$	0
$0.6 \times 2 = 1.2$	1
$0.2 \times 2 = 0.4$	0
$0.4 \times 2 = 0.8$	0
$0.8 \times 2 = 1.6$	1

∴ $(0.3)_{10} = (0.01001...)_2$ **Ans.**

and $(0.6)_{10} = (0.1001...)$. **Ans.**

Example 1.41. *What is hexa-decimal system of number and its field of use? Convert the hexa-decimal number 2AB7 to its binary number.*

Solution: Let us first convert (2AB7)16 to decimal system.

$(2AB7)_{16} = 2 \times 16^3 + 10 \times 16^2 + 11 \times 16^1 + 7 \times 16^0$

$= 8192 + 2560 + 176 + 7$

$= (10935)_{10}$

Now, convert (10935)10 to binary number

		Remainder
2	10935	
2	5467	1
2	2733	1
2	1366	1
2	683	0
2	341	1

2	170	1
2	85	0
2	42	1
2	21	0
2	10	1
2	5	0
2	2	1
2	1	0

$\therefore \quad (2AB7)_{16} = (10101010110111)_2$

Check: $(2AB7)_{16} = (0010\ 1010\ 1011\ 0111)_2$ **Ans.**

Example 1.42. *Convert the decimal 119 into octal and then into binary form.*

Solution: Here radix is 8

8	119	Remainder
8	14	7
	1	6

$\therefore \quad (119)_{10} = (167)_8$

2	119	Remainder
2	59	1
2	29	1
2	14	1
2	7	0
2	3	1
	1	1

$\therefore \quad (119)_{10} = (167)_8 = (1110111)_2$ **Ans.**

Example 1.43. *Find the decimal equivalent of $(1011101.11)_2$ and $(11572.14)_8$.*

Solution:

(*i*) Integer $(1011101)_2 = 1 \times 2^6 + 0 \times 2^5 + 1 \times 2^4 + 1 \times 2^3 + 1 \times 2^2 + 0 \times 2^1 + 1 \times 2^0$

$= 64 + 0 + 16 + 8 + 4 + 0 + 1$

$= (93)_{10}$

Fraction $(0.11)_2 = 1 \times 2^{-1} + 1 \times 2^{-2}$

$= 1/2 + 1/4 = 3/4$

$$= 0.75$$

$\therefore \quad (1011101.11)_2 = (93.75)_{10}$ **Ans.**

(*ii*) Integer $(11572)_8 = 1 \times 8^4 + 1 \times 8^3 + 5 \times 8^2 + 7 \times 8^1 + 2 \times 8^0$

$$= 4096 + 512 + 320 + 56 + 2$$

$$= (4986)_{10}$$

Fraction $(.14)_8 = 1 \times 8^{-1} + 4 \times 8^{-2}$

$$= \frac{1}{8} + \frac{1}{16} = \frac{3}{16}$$

$$= (0.1875)_{10}$$

$\therefore \quad (11572.14)_8 = (4986.1875)_{10}$ **Ans.**

Example 1.44. *(a) Convert the decimal number 0.1 to the binary system.*

(b) Convert the decimal number 5502.8 to the octal system.

(c) Convert the following binary numbers to octal and hexa decimal form:

(i) 110011101 (ii) 11101100111.

Solution:

(*a*) $(0.1)_{10}$ to binary system

	Integer
$0.1 \times 2 = 0.2$	0
$0.2 \times 2 = 0.4$	0
$0.4 \times 2 = 0.8$	0
$0.8 \times 2 = 1.6$	1
$0.6 \times 2 = 1.2$	1
$0.2 \times 2 = 0.4$	0
$0.4 \times 2 = 0.8$	0
$0.8 \times 2 = 1.6$	1
$0.6 \times 2 = 1.2$	1

$\therefore \quad (0.1)_{10} = (000110011...)_2$ **Ans.**

(*b*) $(5502.8)_{10}$ to octal system

Integer			Fraction	
8	5502		$0.8 \times 8 = 0.64$	
8	087	6	$0.64 \times 8 = 5.12$	5
8	85	7	$0.12 \times 8 = 0.96$	0
8	10	5	$0.96 \times 8 = 76.8$	7
	1	2	$(0.8)_{10} = (0.0507)_8$	

$$(5502)_{10} = (12576)_8$$

$\therefore \quad (5502.8)_{10} = (12576.0507)_8$ **Ans.**

(*c*) (*i*) $(110011101)_2$ to octal and hexa decimal systems. $(110011101)_2$, to be first converted to decimal system

i.e.,
$$= 1 \times 2^8 + 1 \times 2^7 + 0 \times 2^6 + 0 \times 2^5 + 1 \times 2 + 0 \times 0^1 + 1 \times 2^0$$
$$= 256 + 128 + 0 + 0 + 16 + 8 + 4 + 0 + 1$$
$$= (413)_{10}$$

Conversion to octal system

		Remainder
8	413	
8	51	5
	6	3

$\therefore \quad (110011101)_2 = (413)_{10} = (635)_8$ **Ans.**

Consersion to hexa-decimal system

		Remainder
16	413	
16	25	13
	1	9

$\therefore \quad (110011101)_2 = (413)_{10} = (19D)_{16}$ **Ans.**

(*ii*) $(11101100111)_2$ to octal and hexa-decimal systems

First convert $(11101100111)_2$ to decimal system

$$(11101100111)_2 = 1 \times 2^{10} + 1 \times 2^9 + 1 \times 2^8 + 0 \times 2^7 + 1 \times 2^6 + 1 \times 2^5 + 0 \times 2^4 + 0 \times 2^3 + 1 \times 2^2 + 1 \times 2^1 + 1 \times 2^0$$
$$= 1024 + 512 + 256 + 0 + 64 + 32 + 0 + 0 + 4 + 2 + 1$$
$$= (1895)_{10}$$

Now

		Remainder
8	1895	
8	236	7
8	29	4
	3	5

$\therefore \quad (11101110011)_2 = (1895)_{10} = (3547)_8$ **Ans.**

Again,

16	1895	
16	118	7
	7	6

$\therefore \quad (11101100111)_2 = (1895)_{10} = (767)_{16}$ **Ans.**

The binary number may also be converted to hexa-decimal system as under:

$(11101100111)_2$ may be grouped into fours from the right

$(0111, 0110,0111)_2$

Now $\quad 0111 = 7, 0110 = 6$ md $011 = 7$

$\therefore \quad (11101100111)_2 = (767)_{16}$ **Ans.**

Example 1.45. *Consider a hexa decimal integer 2D 4AC. Write the number in binary form.*

Solution:

First convert the given number to decimal system

$$(2D\,4AC)_{16} = 2 \times 16^4 + 13 \times 16^3 + 4 \times 16^2 + 10 \times 16^1 + 12 \times 16^0$$
$$= 131072 + 53248 + 1024 + 160 + 12 = (185516)_{10}$$

2	185516	Remainder
2	92758	0
2	46379	0
2	23189	1
2	11594	1
2	5797	0
2	2898	1
2	1449	0
2	724	1
2	362	0
2	181	0
2	90	1
2	45	0
2	22	1
2	11	0
2	5	1
2	2	1
	1	0

$\therefore \quad (185516)_{10} = (101101010010101100)_2$ **Ans.**

Example 1.46. *Obtain the decimal equivalent of $(10101.10101)_2$.*

Solution:

$$(10101.10101)_2 = 1 \times 2^4 + 0 \times 2^3 + 1 \times 2^2 + 0 \times 2^1 + 1 \times 2^0 .\, 1 \times 2^{-1}$$
$$+ 0 \times 2^{-2}\, 2 + 1 \times 2^{-3} + 0 \times 2^{-4} + 1 \times 2^{-5}$$
$$= (16 + 0 + 4 + 0 + 1).\ \frac{1}{2} + 0 + \frac{1}{8} + 0 + \frac{1}{32}$$
$$= (21.65625)_{10} \quad \textbf{Ans.}$$

Example 1.47. *Convert* $(3370.75)_{10}$ *into hexa decimal and convert this hexa decimal number to binary.*

Solution:

		Remainder
16	3370	
16	210	10
	13	2

$$(3370)_{10} = (13, 2, 10)_1 = (D2A)_{16}$$

$$(0.75)_{10}$$

$$.75 \times 16 = 12.0 \qquad 12 = C$$

$$\therefore \quad (3370.75)_{10} = (D2A.C)_{16} \quad \textbf{Ans.}$$

Example 1.48. *Perform the following conversions:*

(*i*) $(512.5)_{10}$ *to binary;*

(*ii*) $(ABCD)_{16}$ *to octal.*

Solution.

(*i*) $(512.5)_{10}$ to binary

Integer Part 512

		Remainder
2	512	
2	256	0
2	128	0
2	64	1
2	32	1
2	16	0
2	8	1
2	16	0
2	8	1
2	4	0
2	2	0
2	1	0

Fraction Part

$$(0.5)_{10} = .5 \times 2 = 1.0$$

i.e., $(0.5)_{10} = (1)_2$

i.e., $(512)_{10} = 1000000000$

$\therefore$ $(512.5)_{10} = (1000000000 + 1)_2 = (100000001)_2$

(*ii*) $(ABCD)_{16}$ to octal system

First convert $(ABCD)_{16}$ to decimal system

i.e., $(ABCD)_{16} = 10 \times 16^3 + 11 \times 16^2 + 12 \times 16^1 + 13 \times 16^0$

$= 40960 + 2816 + 192 + 13$

$= (43981)_{10}$

Now convert $(43981)_{10}$ to octal system

8	43981	Remainder
8	5497	5
8	687	1
8	85	7
8	10	5
	1	2

$\therefore$ $(ABCD)_{16} = (43981)_{10} = (125715)_8$ **Ans.**

Example 1.49. *Convert the decimal equivalent of $(11010.111)_2$, $(736.5)_8$ and $(3FA.8)_{16}$.*

Solution:

(*i*) $(11010.111)_2$ to decimal

(*a*) Integer part

$= (11010)_2$

$= 1 \times 2^4 + 1 \times 2^3 + 0 \times 2^2 + 1 \times 2^1 + 0 \times 2^0$

$= 16 + 8 + 0 + 2 + 0$

$= 26$

(*b*) Fraction Part.

$= 0.111$

$= 1 \times 2^{-1} + 1 \times 2^{-2} + 1 \times 2^{-3}$

$= \frac{1}{2} + \frac{1}{4} + \frac{1}{8}$

$= \frac{7}{8} = .875$

$\therefore$ $(11010.111)_2 = (26.875)_{10}$ **Ans.**

(*ii*) $(736.5)_8$ to decimal

(*a*) Integer Part

$= (736)_8$

$= 7 \times 8^2 + 3 \times 8^1 + 6 \times 8^0$

$= 448 + 24 + 6$

$= 478$

(*b*) Fraction Part

$= 0.5$

$$= .5 \times 8^{-1} = \frac{5}{8} = 0.625$$

$(736.5)_8 = (478.625)_{10}$ **Ans.**

(*iii*) $(3FA.8)_{16}$ to decimal

(*a*) Integer Part $= 3$ F.A.

$= 3 \times 16^2 + 15 \times 16^1 + 10 \times 16^0$

($\therefore$ F = 15 and A = 10)

$= 768 + 240 + 10$

$= 1018$

(*b*) Fraction Part $= 0.8$

$= 8 \times 16^{-1}$

$$= \frac{8}{16} = \frac{1}{2} = 0.5$$

$\therefore$ $(3FA.8)_{16} = (1018.5)$ **Ans.**

Example 1.50. *(i) Convert 567.1875_{10} into hexadecimal.*

(ii) Convert $(A2C)_{16}$ into binary.

(iii) Convert the following 5421 code number with decimal 0010 10001010.

Solution.

(*i*) $(567.1875)_{10}$ into hexadecimal

Integer part =	16	567	Remainder
	16	35	7
		2	3

i.e., $(567)_{10} = (237)_{16}$

Fraction Part $= 0.1875$

$= 0.1875 \times 16 \quad 0.3$

$= 0.3$

$\therefore$ $(567.1875)_{10} = (237.3)_{16}$

(*ii*) $(A.2.C)_{16}$ to binary system

First convert the number into decimal system.

i.e., $(A.2C)_{16}$ to decimal

$= (10.2.12)_{16}$

$= 10 \times 16^2 + 2 \times 16^1 + 12 \times 16^0$

$= 2560 + 32 + 12$

$= (2604)_{10}$

Now convert the decimal number into binary

2	2604	
2	1302	0
2	651	0
2	325	1
2	162	1
2	81	0
2	40	1
2	20	0
2	10	0
2	5	0
2	2	1
	1	0

$\therefore \quad A(A2C)_{16} = (2604)_{10} = (101000101100)_2$ **Ans.**

1.9 NECESSITY OF USING RADIX CONVERSION IN COMPUTERS

All the computers are binary machines. They understand only binary language which is also popularly known as *Machine Language.* This language makes use of only two symbols namely 0 and 1. We may say that the radix (or base) of binary language is 2. But, the data generally processed by computers consists of decimal numbers which have radix (or base) as 10. As a computer does not understand decimal, octal or hexa-decimal number systems, the given data in decimal number system or other number system other than binary system, must be converted to binary system so that the computer can process it. This conversion of data from decimal number system or other number systems, is called *Radix Conversion* and it is a must for computers.

1.10 NORMALIZED FLOATING POINT NUMBERS

In this method of represention, a real number is expressed as a combination of a mantissa and exponent. The mantissa is made less than 1 but greater than or equal to 0.1 and the exponent is the power of 10 which multiplies the mantissa.

For an example, the number 95.65×10^5 is represented 0.9565E7 where E7 is used for 10^7. In this case, 0.9565 is a mantissa and 7 is the exponent. It may be noted that while storing numbers, the leading digit in the mantissa, is always made non-zero by appropriately shifting it and adjusting the value of the exponent, *i.e.*,

$$.003531 \longrightarrow 0.3531 \times 10^{-2} \longrightarrow 0.3531E - 2$$

The shifting of the mantissa to the left till its most significant digit is non-zero, is called *normalization.* The normalization is done to preserve the maximum number of useful digits.

Arithmetic operations with normalized floating point numbers, are done as under:

1. **Addition.** While adding two numbers presented in normalized floating point notation, the exponents of the two numbers, must be made equal and the mantissa shifted appropriately.

Example 1.51. *Add 0.3525E5 to 0.1635E6.*

Solution: We notice that the two exponents are not equal.

Keeping the operand with the larger exponent, as it is, change the mantissa of the operand with smaller exponent right by a number of places equal to the difference in the two exponents. Hence, shift the mantissa of 0.3525E5 right one place. As the arithmetic unit of the CPU of computer accommodates only a 4 digit mantissa, each shift of the mantissa, causes the last digit of the mantissa to be chopped off.

The two operands after the shift are written as under:

$$\begin{array}{r} 0.1635E6 \\ +\ 0.0352E6 \\ \hline 0.1987E6 \end{array}$$

$\therefore$ The sum = 0.1987E6. **Ans.**

Example 1.52. ***Add 0.4543E3 to 0.4675E7.***

Solution: In this case the exponents of two numbers are not equal. Shift the mantissa of the first operand 4 places to the right before adding it to the second operand, *i.e.,*

$$\begin{array}{r} 0.4675E7 \\ 0.0000E7 \\ \hline 1.4675E7 \end{array}$$

$\therefore$ Sum = 0.4675E7.

Example 1.53. *Add 0.7653E4 to 0.6854E4.*

In this case exponents are equal.

The sum of two numbers is as under:

$$\begin{array}{r} 0.7653E4 \\ 0.6854E4 \\ \hline 1.4507E4 \end{array}$$

As the mantissa 1.4507 has 5 digits and is greater than 1, shift the decimal one place. Now the last digit 7 is chopped off.

Hence, answer is 0.1450E4.

Note: It may be noted that the exponent part can not store more than two digits. In case the exponent of the number is more than 99, overflow condition occurs and the arithmetic unit of the computer intimates an error condition.

2. **Subtraction.** The operation of subtraction may be treated as the operation of addition to a negative number as explained in the following examples.

Example 1.54. *Subtract 0.5565E4 from 0.6376E4.*

Solution: As the exponents are equal, the mantissa are subtracted. *i.e.,*

$$\begin{array}{r} 0.6376\text{E}4 \\ -\ 0.5565\text{E}4 \\ \hline 0.0811\text{E}4 \end{array}$$

As the most significant digit of the mantissa is 0, the mantissa must be shifted left till the most significant digit is non-zero. For each left shift of the mantissa, the exponent gets reduced by 1. Hence, the required result is 0.8110E3.

Example 1.55. *Subtract 0.8573-5 from 0.4835E-4.*

Solution: Here, the exponents of both the numbers are not equal. Shift the mantissa with smaller exponent to right and increase the exponent by 1 for each right shift.

$$\begin{array}{r} 0.4835\text{E} - 4 \\ -\ 0.0857\text{E} - 4 \\ \hline = \ 0.3978\text{E} - 4 \end{array} \quad \textbf{Ans.}$$

3. **Multiplication.** While multiplying two numbers in normalized floating point mode, mantissas of two numbers are multiplied and the exponents are added. After multiplying the mantissas, the resultant mantissa is normalized as in the case of addition and subtraction operations and the exponents are adjusted accordingly.

Example 1.56. *Multiply + S435E11 and – .3752 E–14.*

Solution: —.20391 $\boxed{212}$ IE (11 – 14) = .2039E-3. **Ans.** *212 discarded.

Example 1.57. ***Multiply 0.1152E12 × .1235E16.***

Solution: 0.1152E12 × .1235E16

=. 0142272E28 = 0.142272E27

= 0.1422E27. **Ans.**

4. Division. While dividing a number by another, the mantissa of the numerator is divided by the mantissa of the denominator. Also, the exponent of the denominator is subtracted from that of the numerator. The quotient mantissa is normalized to make the most significant digit a non-zero and the resultant exponent adjusted accordingly.

Example 1.58. *Divide 0.2526E5 by 0.8352E3.*

Solution:

$$\frac{0.2526}{0.8352} \text{E}(5-3) = 0.3024\boxed{425}\text{E2}$$

$$= 0.3024\text{E2.} \quad \textbf{Ans.}$$

1.11 INTERNAL REPRESENTATION OF DATA

The internal data in a computer may be represented by (*i*) words and bytes (*ii*) characters. The basic unit of information used for the internal, representation of data in each memory cell, is called a *word.* The most popular word sizes are 24, 30, 32, 36, 48 and 60 bits. Some computers have words of fixed length and some have words of variable length. Each word in the memory is provided its own *address.* All the bits of a word are transmitted simultaneously whenever we provide an input/output statement.

Bits: The smallest unit of information that can be stored and processed by a digital computer is called a *bit.* Binary digits 0 and 1 are the smallest units of information and therefore are called *bits* (a short form of binary digit). In fact, a collection or a sequence of bits is used to represent a unit.

Bytes: The part of the word which can be stored or retrieved from the memory is called a *byte. A* byte normally consists of 6 or 8 bits and may be addressed individually. The computers which provide such a facility are called *byte addressable computers.* For an example, the computers IBM 360/370 series retain words composed of four 8-bit bytes giving a 32 bit word as follows:

Byte 0	Byte 1	Byte 2	Byte 3
8 bits	8 bits	8 bits	8 bits

In byte addressable computers, the address is its left most byte. Address may be defined as a means of locating data or communicating with devices connected to a computer.

1.12 COMPUTER CODES

The binary numbers are coded to represent characters in the computer memory. The most commonly used computer codes are discussed under.

The data to be handled by computers may be classified in the following forms:

1. **Alphanumeric data:** The data which consists of a string of symbols where a symbol may be one of the letters A, B, C, ..., X, Y, Z or one of the digits 0, 1, 2, 9 or a special character + – * /, . ( ) = space or blank, etc., is called *Alphanumeric data.*

2. **Alphabetic data:** The data which consists of only the letters A, B, C, ..., X, Y, Z and the blank character, is called *alphabetic data.*

3. **Numeric data:** The data which consists of only numbers 0, 1, 2, 9 is called *numeric data.*

All the types of data must be represented internally by the bits 0 and 1. This necessitates the binary coding schemes to be used to represent data internally in the computer memory. Representation of every symbol of the data by a group of bits, is known as *binary coding*. The group of bits used to represent a symbol of data is called a *byte*. The byte which contains *n* bits is called *n-bit byte*. An 8-bit byte is mostly used in modern computers.

1. **BCD CODE:** The memory codes generally adopted in early computers was the Binary Coded Decimal (BCD) Code. In this code each digit of a decimal number is converted into its binary equivalent. Refer to the BCD equivalent of each decimal digit shown in Table 1.4.

Table 1.4 BCD Equivalent of Decimal Digits

Decimal Digit	*BCD Equivalent*
0	0000
1	0001
2	0010
3	0011
4	0100
5	0101
6	0110
7	0111
8	1000
9	1001

As decimal digits 8 and 9 require 4 bits to represent each of them, it is customary to represent all decimal digits in BCD by 4 bits.

It may, however, be noted that representation of decimal number in pure binary form is not equal to its value in BCD form.

Example 1.59. *Convert $(52)_{10}$ into pure binary form as well as in BCD form.*

Solution:

2	52	0
2	26	0
2	13	0
2	6	1
2	3	1
	1	1

i.e., $(52)_{10} = (110100)_2$ **Ans.**

(*ii*) For converting $(52)_{10}$ into BCD form, proceed as under:

5 is equivalent to 0101

2 is equivalent to 0010

$$\therefore \quad (52)_{10} = \frac{01010010}{5 \quad 2}$$

or $= 01010010$ in BCD **Ans.**

Advantages and Disadvantages of BCD System: The following are the advantages of BCD system:

(*i*) As each decimal digit is independently converted to a 4 bit binary number, the conversion is very each.

(*ii*) It is only useful to represent decimal numbers.

(*iii*) It fails to accommodate the various characters used by computers.

6-BIT BCD CODE: To overcome the above limitations, computer designers now commonly use 6 bits to represent various characters in BCD code. In this system, the four BCD numeric place positions are retained and two additional zone positions are added on the left.

By adopting 6 bits in BCD code we can represent 64 different characters, *i.e.*, (2^6). This is a sufficient number to code the following characters:

1. Decimal digits 10
2. Alphabetic letters 26
3. Special characters 28

Total = 64

The coding of alphabetic and numeric characters in BCD using 6-bite is shown in Table 1.5.

Table 1.5 Alphabetic and Numeric Characters in BCD

Character	*BCD Code*		*Character*	*BCD Code*	
	Zone	*Digit*		*Zone*	*Digit*
A	11	0001	S	01	0010
B	11	0010	T	01	0011
C	11	0011	U	01	0100
D	11	0100	V	01	0101
E	11	0101	w	01	0110
F	11	0110	X	01	0111
G	11	0111	Y	01	1000
H	11	1000	Z	01	1001
I	11	1001	1	00	0001

Character	BCD Code		Character	BCD Code	
	Zone	Digit		Zone	Digit
J	10	0001	2	00	0010
K	10	0010	3	00	0011
L	10	0011	4	00	0100
M	10	0100	5	00	0101
N	10	0101	6	00	0110
O	10	0110	7	00	0111
P	10	0111	8	00	1000
Q	10	1000	9	00	1001
R	10	1001	0	00	1010

As BCD is a six-bit code, it can be easily divided into two 3-bit groups. 3-bit groups represent 1 octal digit. That's why octal number system is used for memory by computers that use BCD for internal representation of characters.

Example 1.60. *Calculate the binary digits used to record the word DELHI in BCD.*

Solution: By reforming to Table No. 1.5, we get

D = 110100

E = 110101

L = 100011

H = 111000

I = 111001

∴ The binary digits

$$\underbrace{110100}_{D}\quad\underbrace{110101}_{E}\quad\underbrace{100011}_{L}\quad\underbrace{111000}_{H}\quad\underbrace{111001}_{I}$$

record the word DELHI in BCD **Ans.**

Example 1.61. *What binary digits will record the word COMET in BCD ?*

Solution: We know

C = 110011

O = 100110

M = 100100

E = 110101

T = 010011

∴ The binary digits

11011 100110 100100 110101 010011

record the word COMET in BCD. **Ans.**

Example 1.62. *For what statement the binary digits 110111100110010011100110 is used in BCD?*

Solution: The given binary digits are:

110111 100110 010011 100110

Using 6-bit BCD code, we get

110111 100110 010011 100110

Referring to Table No. 1.5, we get

110111 = G

100110 = O

010011 = T

100110 = O

∴ GOTO is the recorded statement in BCD. **Ans.**

2. **EBCDIC:** The BCD code could only represent 64 (2^6) different characters using 6-bit BCD code. This was found insufficient to represent the following:

(*i*) Decimal number	10
(*ii*) Capital letters	26
(*iii*) Lower case letters	26
(*iv*) Special characters	28 +
Total	90 + Characters.

To accommodate 90+ characters, the BCD code was extended from 6-bit code to an 8-bit code. In this system 4 bits are used in the zone. This code which is called the *extended binary-coded decimal interchange code* (EBCDIC) is suitably used to represent 256(2^8) different characters.

As EBCDIC is an 8-bit code, it can be easily divided into two 4-bit groups. Four bit groups represent 1 hexadecimal digit. That's why hexadecimal number system is used for memory by computers that use EBCDIC for internal representation of characters. The alphabetic and numeric characters in EBCDIC are shown in Table 1.6.

Table 1.6 Alphabetic and numeric characters in EBCDIC

Character	*BCD Code*		*Character*	*BCD Code*	
	Zone	*Digit*		*Zone*	*Digit*
A	1100	0001	S	1110	0010
B	1100	0010	T	1110	0011
C	1100	0011	U	1110	0100
D	1100	0100	V	1110	0101
E	1100	0101	W	1110	0110
F	1100	0110	X	1110	0111
G	1100	0111	Y	1110	1000
H	1100	1000	Z	1110	1001
I	1100	1001	0	1111	0000
J	1101	0001	1	1111	0001
K	1101	0010	2	1111	0010
L	1101	0011	3	1111	0011
M	1101	0100	4	1111	0100
N	1101	0101	5	1111	0101
O	1101	0110	6	1111	0110
P	1101	0111	7	1111	0111
Q	1101	1000	8	1111	1000
R	1101	1001	9	1111	1001

Example 1.63. *Using binary notation, write the EBCDIC coding for the word STAR. How many bytes are required for this representation?*

Solution:

S = 11100010 in EBCDIC

T = 11100011 in EBCDIC

A = 11000001 in EBCDIC

R = 11011001 in EBCDIC

∴ The EBCDIC coding for the word STAR in binary notation, is:

$$\underbrace{11100010}_{S}\quad\underbrace{11100011}_{T}\quad\underbrace{11000001}_{A}\quad\underbrace{11011001}_{R}.$$

As each letter requires 8-bits (one byte), total four bytes are needed to represent STAR.

3. **ASCII:** The American Standard Code for Information Interchange (ASCII) is adopted by several American Computer manufacturers for their computer's internal code. This code is mostly used for microcomputers as well as for larger computers.

Types of ASCII: There are two types of ASCII codes.

1. **ASCII-7:** This is a 7-bit code that allows $128(2^7)$ different characters. The first 3-bits are used as zone bits and the last 4-bits indicate the digit. In microcomputers that use 8-bit byte (a group of 8-bits) ASCII-7 is suitably adopted by leaving the leftmost first bit of each byte as a zero.

The alphabetic and numeric characters in ASCII-7 notation are shown in Table 1.7.

Table 1.7 Numeric and alphabetic characters in ASII-7

Character	*ASCII-7 Code*		*Hexa-decimal Equivalent*
	Zone	*Digit*	
0	011	0000	30
1	011	0001	31
2	011	0010	32
3	011	0011	33
4	011	0100	34
5	011	0101	35
6	011	0110	36
7	011	0111	37
8	011	1000	38
9	011	1001	39
A	100	0001	41
B	100	0010	42
C	100	0011	43
D	100	0100	44
E	100	0101	45
F	100	0110	46
G	100	0111	47
H	100	1000	48
I	100	1001	49
J	100	1010	4A
K	100	1011	4B
L	100	1100	4C
M	100	1101	4D
N	100	1110	4E
O	100	1111	4F
P	101	0000	50
Q	101	0001	51
R	101	0010	52
S	101	0011	53
T	101	0100	54
U	101	0101	55
V	101	0110	56
W	101	0111	57
X	101	1000	58
Y	101	1001	59
Z	101	1010	5A

2. **ASII-8:** This is an extended version of ASII-7. It was an 8-bit code that allows 256 (2^8) characters. The zone bit contains 4-bits and the digit contains 4-bits.

Difference between ASCII-7 and ASCII-8:

ASCII-7	*ASCIIS*
1. 7-bit code is used	1. 8-bit code is used.
2. Zone contains 3-bits	2. Zone contains 8-bits.
3. It accomodates 128 characters	3. It accomodates 256 characters.

Example 1.64. *In a micro-computer memory the word SON is stored in binary coding ASCII-7. How many bytes will be required for this representation?*

Solution:

S = 1010011 in ASCII-7 binary code
O = 1001111 in ASCII-7 binary code
N = 1001110 in ASCII-7 binary code

∴ The binary coding for the word SON is: 101001110011111001110

No. of bytes required = 1 × 3 = 3. **Ans.**

1.13 BCD ADDITION AND SUBTRACTION

Whenever a final result of calculation is to be displayed as in calculators, a BCD format is generally preferred to. In this code, each decimal digit 0 to 9 is represented with a 4-bit binary word. In BCD, words are the same as the binary equivalents for 0 through 9.

1. **Addition:** BCD cannot have any digit word with a value greater than 9. Whenever the result of a BCD addition is greater than 9 (1001), a carry is a must.

Moreover, a correction factor of 6 must be added to the result if the result in the lower 4-bits is greater than 9 or if the initial addition produces a carry of bit 3 of any BCD digit word. Such a correction, is known as *decimal adjust operation.*

The necessity of this correction factor of 6 is because of the fact that in BCD we need a carry into the next digit after 1001 or 9. But in binary a carry of the lower four bits does not occur until after 1111 or 15 which is definitely more than 9 by 6.

Example 1.65. *Show additions of*

(i) 25_{10} *and* 23_{10} *in BCD*
(ii) 7_{10} *and* 6_{10} *in BCD*
(iii) 8_{10} *and* 9_{10} *in BCD*

Solution:

		BCD	
(i)	25	00100101	
	+ 23	+ 00100011	
	48	01001000	**Ans.**

		BCD	
(ii)	7	0111	
	+ 6	+ 0110	
	13	1101	incorrect BCD because result is more than 9.
		110	Add correction factor of 6
		00010011	correct BCD 13 **Ans.**

		BCD	
(iii)	8	1000	
	+ 9	+ 1001	
	17	10001	incorrect BCD.
		110	Add correction factor of 6
		00010111	correct BCD 17 **Ans.**

2. **Subtraction:** BCD subtraction may be carried out in the same way as biliary subtraction. In case the final result is greater than 1001, or if a borrow is taken from the next higher digit, then a correction factor of 6 is subtracted from the BCD digit word of the final result.

Example 1.66. *Perform the operation of subtraction in BCD code in the following case:*

(i) 8_{10} *from* 18_{10}

(ii) 5_{10} *from* 12_{10}

(iii) 9_{10} *from* 16_{10}

Solution:

		BCD	
(i)	18	00011000	
	− 7	− 0111	
	11	00010001	a legal BCD code = 11_{10} **Ans.**

		BCD	
(ii)	12	10010	
	− 4	− 0100	
	8	01110	not a legal BCD code
		(−) 110	subtract 6
		1000	= 8_{10} **Ans.**

		BCD	
(*iii*)	16	1 0110	
	– 9	– 0 1001	
	7	0 1101	not a legal BCD code,
		(–) 110	Subtract 6
		0111	= 7_{10} **Ans.**

Example 1.67. *Convert the following decimal numbers to B.C.D. number.*

(i) + 135

(ii) – 327

Solution: Referring to ECDIC table 1.5, we get the required conversions as under

	1	3	5	+
(*i*)	0001	0011	0101	1100

i.e., $(135)_{10}$ = [0001001101011100] **Ans.**

	3	2	7	–
(*ii*)	0011	0010	0111	1101

i.e., $(327)_{10}$ = [011001001111101] **Ans.**

1.14 COMPLEMENTARY SUBTRACTION

The nine's complements. The nine's complements of a decimal number may be obtained by subtracting each decimal digit from 9.

The ten's complements. The ten's complement of a number is its nine's complement plus one.

Illustration. Calculate the nine's complements and ten's complements of decimal numbers 125, 325, 450, 725.

Decimal number	125	325	450	725
Nine's complement	874	674	549	274
Ten's complement	875	675	550	275

In case of binary numbers, the corresponding complements are called *one's complement* and *two's complement* respectively.

The one's complements. The 1's complement of a binary number may be obtained by subtracting each binary digit form 1.

Illustration. Calculate the one's complement and two's complement of the binary numbers 11001, 10101, 101101.

Binary number	11001	10101	101101
One's complement	00110	01010	010010
Two's complement	00111	01011	010011

The 2's Complementary. The 2's complement of a binary number is its l's complement plus 1.

1. **Subtraction by 9's complement.** Subtraction by 9's complement method may be done with the following steps:

(*i*) Find the 9's complement of the number which is to be subtracted.

(*ii*) Add this complement to the number from which number is to be subtracted.

(*iii*) If there is a carry of 1, add it to obtain the required result.

(*iv*) If there is no carry of 1, add 0. Recomplement it and attach a negative sign to obtain the desired result.

Example 1.68. *Perform the following operation 45-15 using 9's complement.*

Solution:

Normal subtraction	9's complement method	
45	45	
− 15	+ 84	(Add 9's complement of 15)
30	[1] 29	
	1	(Add the carry of 1)
	Result = 30	

Example 1.69. *Perform the following operation 23-75 using 9's complement method.*

Solution:

Normal subtraction	9's complement method	
23	23	
− 75	+ 24	(Add 9's complement of 75)
− 52	47	
	0	(Add 0 as no carry)
	47	
	Result = − 52	(Recomplement and attach a negative sign)

It may be mentioned here that performing a decimal subtraction by 9's complement method, is a tedious task. But, in the case of binary number system, it is very easy because it contains simple several of bits of the binary number. For example, 1's complement of 10101 is 01010.

The method of subtraction by l's complement method in binary number system, is explained in the following solved examples.

Example 1.70. *Subtract 1111 from 101101 by using 1's complement method.*

Solution:

Normal subtraction		9's complement method	
101101	(45)	101101	
– 001111	(15)	+ 110000	(Add 1's complement of
		[1] 011101	001111)
		1	(Add the carry of 1)
11110	(30)	11110	Ans.

Example 1.71. *Subtract 1001011 from 10111 by using 1's complement method.*

Solution:

Normal binary subtraction		Complementary method	
10111	(23)	0010111	
–1001011	(75)	+ 0110100	(Add 1's complement of
–110100	–52		1001011)
		1001011	
		+ 0	(Add 0 as no carry)
–110100	–52	1001011	(Recomplement and attach a negative sign)
		–110100	Ans.

2. **Subtraction by 10's Complement.** Subtraction by 10's complementary method may be done with the following steps:

(*i*) Find the 10's complement of the number to be subtracted.

(*ii*) Add this complement to the number from which subtracting.

(*iii*) If there is a carry of 1, ignore it to obtain the required result. If there is no carry of 1, add 0. Again, find 10's complement and attach a negative sign to obtain the result.

Example 1.72. *Perform the following operations.*

(*i*) *24-16 using 10's complement.*

(*ii*) *125-375 using 10's complement.*

Solution:

(*i*) Normal subtraction

```
  24
- 16
----
   8
```

10's Complement method

```
    24
  + 84    (Add 10's complement of 16)
------
[1] 08
     8    (ignoring the carry over)
------
          Ans.
```

(*ii*) Normal subtraction

```
  125
- 375
-----
  250
```

10's Complement method

```
  125
+ 625   (Add 10's complement of 375)
-----
  750
 + 10   (Add 0 as no carry over)
-----
  750
- 250   (Find 10's complement of 750 and attach a negative sign) Ans.
-----
```

3. **Subtraction by 2's Complement.** Subtraction by 2's complement method may be done in the following steps:

(*i*) Find the 2's complement of the number to be subtracted.

(*ii*) Add this complement to the number from which subtracting.

(*iii*) If there is a carry of 1, ignore it to obtain the required number.

(*iv*) If there is no carry of 1, add 0. Find 2's complement and attach a negative sign to obtain the required result.

Example 1.73. *Perform the following operations*

(i) 11101-1101 using 2's complement

(ii) 1010-1011 using 2's complement

Solution:

(*i*) Normal subtraction

```
  11101
- 01101
-------
  10000
```

2's Complement method

```
     11101
   + 10011   (Add 2's complement of 1101)
----------
[1]  10000
     10000   (Ignore the Carry of 1)  Ans.
```

(*ii*) Normal subtraction 2's Complement method

Normal subtraction	2's Complement method	
1010	1010	
– 1011	+ 0101	(Add 2's complement of 1011)
1	1111	
	– 0001	(Find 2's complements and attach a negative sign.) **Ans.**

Floating point numbers. While dividing an integer by an integer, we get the result also in integer but its fractional part is not stored. That's why floating point numbers are used.

Every real number can be normalized to yield a unique sign, a unique binary fraction and a unique binary exponent. Hence, a real number can be stored as follows:

Sign	Exponent	Fraction
1	7	24

For a 32 bit word, one bit is reserved for the sign, 7 bits for the exponent and 24 bits for the binary fraction.

To avoid the use of minus sign in the exponent within its 7 bit field, the true exponent may be converted to its so called characteristic by adding $(100000)_2 = 2^6 = 64$ and then the characteristic is stored in lieu of the true exponent.

The true exponents and their characteristics are shown under:

True exponent: –64 – 63 – 62... – 60 62 63

Characteristic: 012...6364...126...127

In case a 8 digit field is used for the exponent, we add $2^7 = 128$ to the true exponent to obtain its characteristic.

Illustration. Consider a real number – 45.75

Binary number = – 101101.11 Normalized binary number

$$= -0.010110111 \times 2^6$$

Here the exponent is 6 and so the characteristic is 6 + 64 = 70 = (1000110)2.

1.15 REPRESENTATION OF REAL NUMBERS AS POINTS ON A LINE

The points on a straight line are used to represent real numbers. The geometrical representation of real numbers has been found very useful as explained under:

Suppose X'X to be part of a straight line. Mark two points O and A on the straight line X'X such that A is to the right of O. Thus point O divides the given line X'X in two parts, one part to the right of point O called positive and the other part to the left of O called negative. A straight line for which positive and negative sides are fixed, is called a directed line (Fig. 1.1).

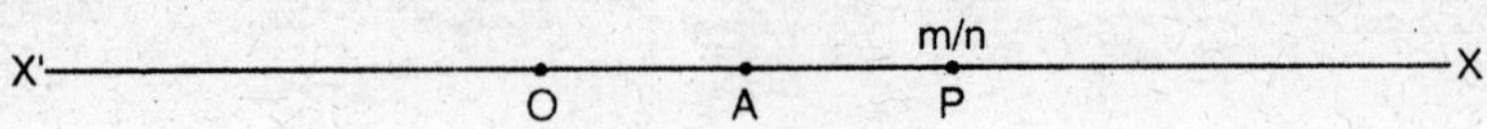

Fig. 1.1

Let the points O and A represent rational numbers zero and 1 respectively such that the distance OA is unity on a certain scale. To represent a rational number $\frac{m}{n}$ $(n > 0)$, take a point P on the right of O if m is positive and to the left of O if m is negative. Thus, OP is m times the nth part of the unit length OA. The point P coincides with O if m is zero. The point P apparently represents the rational number $\frac{m}{n}$. We may thus say that any rational number can be made to correspond to a point of the straight line. If we call the points on the straight line on rational numbers, then there can be imagined infinite number of rational points lie between any two different rational points i.e., *between any two rational numbers, there lie infinitely many rational numbers.*

❑❑❑

2 Determinants

2.1 INTRODUCTION

Determinants are used to handle complicated expressions which occur in electrical and mechanical engineering. Determinants are also advantageously employed in solving algebraic equations.

Let us consider the following **two** simullaneous equations.

$$a_1x + b_1y = 0 \quad \text{...(1)}$$

$$a_2x + b_2y = 0 \quad \text{...(2)}$$

By multiplying equation (1) by b_2 and equation (2) by b_1 we get

$$a_1b_2x + b_1b_2y = 0 \quad \text{...(3)}$$

$$a_2b_1x + b_1b_2y = 0 \quad \text{...(4)}$$

Subtracting eqn. (4) from eqn. (3) we get

$$(a_1b_2 - a_2b_1)x = 0$$

By diving the equation by x we get

$$a_1b_2 - a_2b_1 = 0 \quad \text{...(5)}$$

Equation (5) may be written as

$$\begin{vmatrix} a_1 & b_1 \\ a_2 & b_2 \end{vmatrix} = 0 \quad \text{...(6)}$$

The expression on the left of the equation (6) is called a *determinant*. The members comprising the rows and columns of a determinant, are called its *elements* or *continents*.

2.2 ORDER OF DETERMINANTS

The order of the determinant is decided by the number of rows and number of columns it contains. If it contains of one row and one column, it is said to be of the first order. If it consists of two rows and two columns and in its expanded form each term is a product of two quantities, it is said to be of the second order. If it consists of three rows and three columns, and in its expanded form, each term is a product of three quantities, it is said to be of the third order. Hence, a determinant having n rows and n columns is said to be of nth order.

Examples. 2.1. *Find the order of the following determinants:*

(*i*) $\begin{vmatrix} 5 & 3 \\ 4 & 2 \end{vmatrix}$ (*ii*) $\begin{vmatrix} a & b \\ m & n \end{vmatrix}$

(*iii*) $\begin{vmatrix} 1 & b \\ a & 1 \end{vmatrix}$ (*iv*) $\begin{vmatrix} 2 & 3 & 5 \\ 1 & 5 & 3 \\ 4 & 1 & 5 \end{vmatrix}$

(*v*) $\begin{vmatrix} a & b & c \\ e & f & g \\ h & k & l \end{vmatrix}$

Solution:

(*i*) The given determinant $\begin{vmatrix} 5 & 3 \\ 4 & 2 \end{vmatrix}$ contains two rows and two columns.

∴ It is of the second order. **Ans.**

(*ii*) The given determinant $\begin{vmatrix} a & b \\ m & n \end{vmatrix}$ contains two rows and two columns.

∴ It is of the second order. **Ans.**

(*iii*) The given determinant $\begin{vmatrix} 1 & b \\ a & 1 \end{vmatrix}$ contains rows and two columns.

∴ It is of the second order. Ans.

(*iv*) The given determinant $\begin{vmatrix} 2 & 3 & 5 \\ 1 & 5 & 3 \\ 4 & 1 & 5 \end{vmatrix}$ contains three rows and three columns.

∴ It is of the third order. **Ans.**

(*v*) The given determinant $\begin{vmatrix} a & b & c \\ e & f & g \\ h & k & l \end{vmatrix}$ contains three rows and three columns.

∴ It is of the third order. Ans.

2.3 EXPANDED FORM OF A DETERMINANT

The expansion of a determinant is done as order:

(*a*) **Second Order Determinant.**

$$\begin{vmatrix} a & b \\ c & d \end{vmatrix} = ad - bc$$

To obtain the expanded form of a determinant we take the following steps:

1. Multiply the first constituent of the first row first column and the second constituent of the second row and second column and write as the first term of the expanded form.
2. Use a negative sign.
3. Multiply the second constituent of the first row or first constituent of the second column and the first constituent of second rows or second constituent of the first column and write the next term of the expanded form.

(*b*) **Third Order Determinants:**

$$\begin{vmatrix} a_1 & b_1 & c_1 \\ a_2 & b_2 & c_2 \\ a_3 & b_3 & c_3 \end{vmatrix}$$

To obtain the expanded form take the following steps:

$$\begin{vmatrix} a_1 & b_1 & c_1 \\ a_2 & b_2 & c_2 \\ a_3 & b_3 & c_3 \end{vmatrix} = a_1 \begin{vmatrix} b_2 & c_2 \\ b_3 & c_3 \end{vmatrix} - b_1 \begin{vmatrix} a_2 & c_2 \\ a_3 & c_3 \end{vmatrix} + c_1 \begin{vmatrix} a_2 & b_2 \\ a_3 & b_3 \end{vmatrix}$$

$$= a_1 (b_2c_3 - c_2b_3) - b_1 (a_2c_3 - c_2a_3) + c_1 (a_2b_3 - b_2a_3)$$

Explanation: Multiply the first element of the determinant of the 1st row by the second order determinant obtained by eliminating the first row and first column of the given determinant.

After putting a negative sign, multiply the second element of the first row by the second order determinant obtained by eliminating the first row and second column of the third order determinant.

After putting a positive sign, multiply the third element of the first row by the second order determinant obtained by eliminating the first row and the third column of the given 3rd order determinant.

Note: The arithmetic sign after each product is entered alternatively, i.e., +, –, +, –, ...

It may be noted that a determinant is simply an algebraic expression written in a convenient and concise form to suit the eliminating x and y of simultaneous equations. The elements in the determinant are taken in the same order as in the given equations.

Example. 2.2. *Find the value of the determinant*

$$\begin{vmatrix} 1 & -3 & 1 & 0 \\ 2 & 1 & 0 & -1 \\ 3 & -2 & -1 & -2 \\ 4 & -1 & 0 & 3 \end{vmatrix}$$

Solution.

$$\begin{vmatrix} 1 & -3 & 1 & 0 \\ 2 & 1 & 0 & -1 \\ 3 & -2 & -1 & -2 \\ 4 & -1 & 0 & 3 \end{vmatrix}$$

$$= 1\begin{vmatrix} 1 & 0 & -1 \\ -2 & -1 & -2 \\ -1 & 0 & 3 \end{vmatrix} + 3\begin{vmatrix} 2 & 0 & -1 \\ 3 & -1 & -2 \\ 4 & 0 & 3 \end{vmatrix}$$

$$+ 1\begin{vmatrix} 2 & 1 & -1 \\ 3 & -2 & -2 \\ 4 & -1 & 3 \end{vmatrix} - 0\begin{vmatrix} 2 & 1 & 0 \\ 3 & -2 & -1 \\ 4 & -1 & 0 \end{vmatrix}$$

$$= \begin{vmatrix} 1 & 0 & -1 \\ -2 & -1 & -2 \\ -1 & 0 & 3 \end{vmatrix} + 3\begin{vmatrix} 2 & 0 & -1 \\ 3 & -1 & -2 \\ 4 & 0 & 3 \end{vmatrix} + \begin{vmatrix} 2 & 1 & -1 \\ 3 & -2 & -2 \\ 4 & -1 & 3 \end{vmatrix}$$

$$= 1\begin{vmatrix} -1 & -2 \\ 0 & 3 \end{vmatrix} - 0\begin{vmatrix} -2 & -2 \\ -1 & 3 \end{vmatrix} - 1\begin{vmatrix} -2 & -1 \\ -1 & 0 \end{vmatrix}$$

$$+ \left[6\begin{vmatrix} -1 & -2 \\ 0 & 3 \end{vmatrix} - 0\begin{vmatrix} 3 & -2 \\ 4 & 3 \end{vmatrix} - 3\begin{vmatrix} 3 & -1 \\ 4 & 0 \end{vmatrix}\right]$$

$$+ 2\begin{vmatrix} -2 & -2 \\ -1 & 3 \end{vmatrix} - 1\begin{vmatrix} 3 & -2 \\ 4 & -3 \end{vmatrix} + (-1)\begin{vmatrix} 3 & -2 \\ 4 & -1 \end{vmatrix}$$

$$= 1(-3+0) + 0 + (-1)(0-1) + 6(-3+0) + 0 - 3$$

$$(0+4) + 2(-6-2) - 1(9+8) - 1(-3+8)$$

$$= -3 + 0 + 1 - 18 + 0 - 12 - 16 - 17 - 5 = -70$$

2.4 PROPERTIES OF DETERMINANTS

The main properties of determinants are discussed with reference to a third order determinant.

1. *The value of a determinant is not altered by changing its columns into rows and its rows into columns.*

i.e.,
$$\begin{vmatrix} a_1 & b_1 & c_1 \\ a_2 & b_2 & c_2 \\ a_3 & b_3 & c_3 \end{vmatrix} = \begin{vmatrix} a_1 & a_2 & a_3 \\ b_2 & b_2 & b_3 \\ c_1 & c_2 & c_3 \end{vmatrix}$$

Proof. L.H.S. determinant

$$\begin{vmatrix} a_1 & b_1 & c_1 \\ a_2 & b_2 & c_2 \\ a_3 & b_3 & c_3 \end{vmatrix} = a_1 \begin{vmatrix} b_2 & c_2 \\ b_3 & c_3 \end{vmatrix} - b_1 \begin{vmatrix} a_2 & c_2 \\ a_3 & c_3 \end{vmatrix} + c_1 \begin{vmatrix} a_2 & b_2 \\ a_3 & b_3 \end{vmatrix}$$

$$= a_1 (b_2c_3 - b_3c_2) - b_1 (a_2c_3 - a_3c_2) + c_1 (a_2b_3 - a_3b_2)$$
$$= a_1 (b_2c_3 - b_3c_2) - a_2b_1c_3 + a_3b_1c_2 + a_2b_3c_1 - a_3b_2c_1$$
$$= a_1 (b_2c_3 - b_3c_2) - a_2 (c_3b_1 - b_3c_1) + a_3 (b_1c_2 - b_2c_1)$$

= The R.H.S. determinant.

2. *If two adjacent rows or columns of a determinant are interchanged, the sign of the determinant is only changed and its numerical value remains unaltered.*

Proof. Let the given determinant be

$$\begin{vmatrix} a_1 & a_2 & a_3 \\ b_1 & b_2 & b_3 \\ c_1 & c_2 & c_3 \end{vmatrix}$$

By interchanging the first and second columns, we get

$$\begin{vmatrix} a_2 & a_1 & a_3 \\ b_2 & b_1 & b_3 \\ c_2 & c_1 & c_3 \end{vmatrix}$$

i.e.,
$$\begin{vmatrix} a_1 & a_2 & a_3 \\ b_1 & b_2 & b_3 \\ c_1 & c_2 & c_3 \end{vmatrix} = - \begin{vmatrix} a_2 & a_1 & a_3 \\ b_2 & b_1 & b_3 \\ c_2 & c_1 & c_3 \end{vmatrix}$$

Now, considering the R.H.S. determinant, we get

$$- \begin{vmatrix} a_2 & a_1 & a_3 \\ b_2 & b_1 & b_3 \\ c_2 & c_1 & c_3 \end{vmatrix} = a_2 (b_1c_3 - c_1b_3) - a_1 (b_2c_3 - c_2b_3) + a_3 (b_2c_1 - b_1c_2)$$

$$= a_2 (b_1c_3 - c_1b_3) + a_1 (b_2c_3 - c_2b_3) + a_3 (b_2c_1 - b_1c_2)$$
$$= a_1 (b_2c_3 - c_2b_3) - a_2 (b_1c_3 - c_1b_3) + a_3 (b_1c_2 - b_2c_1)$$
$$= a_1 \begin{vmatrix} b_2 & b_3 \\ c_2 & c_3 \end{vmatrix} - a_2 \begin{vmatrix} b_1 & b_3 \\ c_1 & c_3 \end{vmatrix} + a_3 \begin{vmatrix} b_2 & b_2 \\ c_1 & c_2 \end{vmatrix}$$

$$= \begin{vmatrix} a_1 & a_2 & a_3 \\ b_1 & b_2 & b_3 \\ c_1 & c_2 & c_3 \end{vmatrix} = \text{L.H.S. determinant}$$ **Proved.**

3. *If two rows or two columns of a determinant are identical, the value of the determinant is zero, i.e.,*

$$\begin{vmatrix} a_1 & b_1 & c_1 \\ a_2 & b_2 & c_2 \\ a_3 & b_3 & c_3 \end{vmatrix} = 0$$

Proof. Let the value of the determinant on the left side be Δ.

By changing the first and second columns, we get the value of the determinant as $-\Delta$.

But the two determinants are numerically equal

$\therefore$ $\quad \Delta = -\Delta$

or $\quad \Delta = 0$ **Proved.**

4. *If each element in any row or any column is multiplied by the same factor, then the determinant is multiplied by that factor.*

Let a 3rd order determinant

$$\Delta = \begin{vmatrix} a_1 & b_1 & c_1 \\ a_2 & b_2 & c_2 \\ a_3 & b_3 & c_3 \end{vmatrix} = a_1A_1 + a_2A_2 + a_3A_3$$

Multiplying the first column by K, we get

$$\begin{vmatrix} Ka_1 & b_1 & c_1 \\ Ka_2 & b_2 & c_2 \\ Ka_3 & b_3 & c_3 \end{vmatrix} = Ka_1A_1 + Ka_2A_2 + Ka_3A_3$$

$$= K(a_1A_1 + a_2A_2 + a_3A_3)$$

$$= K\Delta.$$

Cor. If each constituent of one row or one column, is the same multiple of another row, or column the determinant vanishes.

5. *If each element in any row or a column, consists of two terms, then the determinant is expressed as the sum of two other determinants.*

Proof. Consider the following determinant:

$$\begin{vmatrix} a_1 + d_1 & b_1 & c_1 \\ a_2 + d_2 & b_2 & c_2 \\ a_3 + d_3 & b_3 & c_3 \end{vmatrix}$$

Expanding the determinant column rise, we get

$(a_1 + d_1)A_1 - (a_2 + d_2)A_2 + (a_3 + d_3)A_3$

$$= a_1A_1 + d_1A_1 - a_2A_2 + d_2A_2 + a_3A_3 + d_3A_3$$
$$= (a_1A_1 + a_2A_2 - a_3A_3) + (d_1A_1 - d_2A_2 + d_3A_3)$$
$$= \begin{vmatrix} a_1 & b_1 & c_1 \\ a_2 & b_2 & c_2 \\ a_3 & b_3 & c_3 \end{vmatrix} + \begin{vmatrix} d_1 & b_1 & c_1 \\ d_2 & b_2 & c_2 \\ d_3 & b_3 & c_3 \end{vmatrix}$$

$$\therefore \begin{vmatrix} a_1+d_1 & b_1 & c_1 \\ a_2+d_2 & b_2 & c_2 \\ a_3+d_3 & b_3 & c_3 \end{vmatrix} = \begin{vmatrix} a_1 & b_1 & c_1 \\ a_2 & b_2 & c_2 \\ a_3 & b_3 & c_3 \end{vmatrix} + \begin{vmatrix} d_1 & b_1 & c_1 \\ d_2 & b_2 & c_2 \\ d_3 & b_3 & c_3 \end{vmatrix}$$ **Proved.**

Similarly it can be shown that

$$\begin{vmatrix} a_1+d_1+m_1 & b_1 & c_1 \\ a_2+d_2+m_2 & b_2 & c_2 \\ a_3+d_3+m_3 & b_3 & c_3 \end{vmatrix} = \begin{vmatrix} a_1 & b_1 & c_1 \\ a_2 & b_2 & c_2 \\ a_3 & b_3 & c_3 \end{vmatrix} + \begin{vmatrix} d_1 & b_1 & c_1 \\ d_2 & b_2 & c_2 \\ d_3 & b_3 & c_3 \end{vmatrix} + \begin{vmatrix} m_1 & b_1 & c_1 \\ m_2 & b_2 & c_2 \\ m_3 & b_3 & c_3 \end{vmatrix}$$

Example. 2.3. *Prove that:*

$$\begin{vmatrix} a_1+\alpha_1 & b_1+\beta_1 & c_1 \\ a_2+\alpha_2 & b_2+\beta_2 & c_2 \\ a_3+\alpha_3 & b_3+\beta_3 & c_3 \end{vmatrix} = \begin{vmatrix} a_1 & b_1 & c_1 \\ a_2 & b_2 & c_2 \\ a_3 & b_3 & c_3 \end{vmatrix} + \begin{vmatrix} a_1 & \beta_1 & c_1 \\ a_2 & \beta_2 & c_2 \\ a_3 & \beta_3 & c_3 \end{vmatrix}$$
$$+ \begin{vmatrix} \alpha_1 & b_1 & c_1 \\ \alpha_2 & b_2 & c_2 \\ \alpha_3 & b_3 & c_3 \end{vmatrix} + \begin{vmatrix} \alpha_1 & \beta_1 & c_1 \\ \alpha_2 & b_2 & c_2 \\ \alpha_3 & \beta_3 & c_3 \end{vmatrix}$$

Solution: L.H.S. $= \begin{vmatrix} a_1+\alpha_1 & b_1+\beta_1 & c_1 \\ a_2+\alpha_2 & b_2+\beta_2 & c_2 \\ a_3+\alpha_3 & b_3+\beta_3 & c_3 \end{vmatrix}$

$$= \begin{vmatrix} a_1 & b_1+\beta_1 & c_1 \\ a_2 & b_2+\beta_2 & c_2 \\ a_3 & b_3+\beta_3 & c_3 \end{vmatrix} + \begin{vmatrix} \alpha_1 & b_1+\beta_1 & c_1 \\ \alpha_2 & b_2+\beta_2 & c_2 \\ \alpha_3 & b_3+\beta_3 & c_3 \end{vmatrix}$$

$$= \begin{vmatrix} a_1 & b_1 & c_1 \\ a_2 & b_2 & c_2 \\ a_3 & b_3 & c_3 \end{vmatrix} + \begin{vmatrix} a_1 & \beta_1 & c_1 \\ a_2 & \beta_2 & c_2 \\ a_3 & \beta_3 & c_3 \end{vmatrix}$$
$$+ \begin{vmatrix} \alpha_1 & b_1 & c_1 \\ \alpha_2 & b_2 & c_2 \\ \alpha_3 & b_3 & c_3 \end{vmatrix} + \begin{vmatrix} \alpha_1 & \beta_1 & c_1 \\ \alpha_2 & b_2 & c_2 \\ \alpha_3 & \beta_3 & c_3 \end{vmatrix}$$ **Proved.**

Example. 2.4. *Find the value of*

$$\begin{vmatrix} 64 & 18 & 22 \\ 36 & 12 & 15 \\ 63 & 21 & 23 \end{vmatrix}$$

Solution:

By splitting the first column, we get

$$\begin{vmatrix} 64 & 18 & 22 \\ 36 & 12 & 15 \\ 71 & 21 & 23 \end{vmatrix} = \begin{vmatrix} \overbrace{10 \quad 54} & 18 & 22 \\ 0 \quad 36 & 12 & 15 \\ 8 \quad 63 & 21 & 23 \end{vmatrix}$$

$$= \begin{vmatrix} 10 & 18 & 22 \\ 0 & 12 & 15 \\ 8 & 21 & 23 \end{vmatrix} + \begin{vmatrix} 54 & 18 & 22 \\ 36 & 12 & 15 \\ 63 & 21 & 23 \end{vmatrix}$$

$$= \begin{vmatrix} 10 & 18 & 22 \\ 0 & 12 & 15 \\ 8 & 21 & 23 \end{vmatrix} + \begin{vmatrix} 10 & 18 & 18+4 \\ 0 & 12 & 12+3 \\ 8 & 21 & 21+2 \end{vmatrix}$$

$$= \begin{vmatrix} 10 & 18 & 18 \\ 0 & 12 & 12 \\ 8 & 21 & 21 \end{vmatrix} + \begin{vmatrix} 10 & 18 & 4 \\ 0 & 12 & 3 \\ 8 & 21 & 2 \end{vmatrix}$$

$$= 10\,[24 - 63]\;0\,[18 \times 2 - 84] + 8\,[54 - 48]$$

$$- 10 \times 39 + 8 \times 6 = -390 + 48 = -342$$ **Ans.**

6. *By increasing or diminishing each element of a row or a column by the same multiple of the corresponding element of another row or column, the value of the determined remains unaltered, i.e.,*

$$\begin{vmatrix} a_1 & b_1 & c_1 \\ a_2 & b_2 & c_2 \\ a_3 & b_3 & c_3 \end{vmatrix} = \begin{vmatrix} a_1 + kb_1 & b_1 & c_1 \\ a_2 + kb_2 & b_2 & c_2 \\ a_3 + kb_3 & b_3 & c_3 \end{vmatrix}$$

where k many any positive or negative integer.

Proof. Consider the determinant on the right hand side,

i.e.,
$$\begin{vmatrix} a_1 + kb_1 & b_1 & c_1 \\ a_2 + kb_2 & b_2 & c_2 \\ a_3 + kb_3 & b_3 & c_3 \end{vmatrix} = \begin{vmatrix} a_1 & b_1 & c_1 \\ a_2 & b_2 & c_2 \\ a_3 & b_3 & c_3 \end{vmatrix} + \begin{vmatrix} kb_1 & b_1 & c_1 \\ kb_2 & b_2 & c_2 \\ kb_3 & b_3 & c_3 \end{vmatrix}$$

or
$$\begin{vmatrix} a_1+kb_1 & b_1 & c_1 \\ a_2+kb_2 & b_2 & c_2 \\ a_3+kb_3 & b_3 & c_3 \end{vmatrix} = \begin{vmatrix} a_1 & b_1 & c_1 \\ a_2 & b_2 & c_2 \\ a_3 & b_3 & c_3 \end{vmatrix}$$

= L.H.S. determinant. **Proved.**

Example. 2.5. *Find the value of*

$$\begin{vmatrix} 4 & 7 & 8 & 12 \\ 2 & 2 & 4 & 6 \\ 1 & 3 & 4 & 5 \\ 4 & 3 & 10 & 18 \end{vmatrix}$$

Solution:

$$\begin{vmatrix} 4 & 7 & 8 & 12 \\ 2 & 2 & 4 & 6 \\ 1 & 3 & 4 & 5 \\ 4 & 5 & 10 & 18 \end{vmatrix} = \begin{vmatrix} 4 & 3 & 0 & 0 \\ 2 & 0 & 0 & 0 \\ 1 & 2 & 2 & 2 \\ 4 & 1 & 2 & 6 \end{vmatrix} \begin{matrix} c_2 = c_2 - c_1 \\ c_3 = c_3 - 2c_1 \\ c_4 = c_4 - 3c_1 \end{matrix}$$

$$= 4\begin{vmatrix} 0 & 0 & 0 \\ 2 & 2 & 2 \\ 1 & 2 & 6 \end{vmatrix} - 3\begin{vmatrix} 2 & 0 & 0 \\ 1 & 2 & 2 \\ 4 & 2 & 6 \end{vmatrix}$$

$$= -3\,(\,2\,(12-4)\,) = -3 \times 16 = -48$$ **Ans.**

Example. 2.6. *Show that*

$$\begin{vmatrix} 1 & 1 & 1 \\ l & m & n \\ l^3 & m^3 & n^3 \end{vmatrix} = (m-n)(n-l)\,(l-m)\,(l+m+n)$$

Solution:

$$\begin{vmatrix} 1 & 1 & 1 \\ l & m & n \\ l^3 & m^3 & n^3 \end{vmatrix}$$

$$c_2 = c_2 - c_1$$
$$c_2 = c_2 - c_1$$

$$= \begin{vmatrix} 1 & 0 & 0 \\ l & n-l & n-l \\ l^3 & m^3-l^3 & n^3-l^3 \end{vmatrix} = 1\begin{vmatrix} m-l & n-l \\ m^3-l^3 & n^3-l^3 \end{vmatrix}$$

$$= (m-l)(n-l)\begin{vmatrix} 1 & 1 \\ m^2+ml+l^3 & n^2+nl+l^2 \end{vmatrix}$$

$$= (m-l)(n-l)(n^2+nl+l^2-m^2-ml-l^2)$$
$$= (m-l)(n-l)(n^2-m^2+l(n-m))$$
$$= (m-l)(n-l)(n-m)(n+m+l)$$
$$= n^2-m^2+l(n-m)(n-m)$$
$$= (m-n)(l-m)(n-l)(l+m+n)$$

Proved.

Example. 2.7. *Prove that*

$$\begin{vmatrix} 1 & a & a^2 \\ 1 & b & b^2 \\ 1 & c & c^2 \end{vmatrix} = (b-c)(c-a)(a-b)$$

Solution: The given determinant is

$$\begin{vmatrix} 1 & a & a^2 \\ 1 & b & b^2 \\ 1 & c & c^2 \end{vmatrix}$$

By changing the rows into columns, we get

$$\begin{vmatrix} 1 & 1 & 1 \\ a & b & c \\ a^2 & b^2 & c^2 \end{vmatrix} = \begin{vmatrix} 1 & 0 & 0 \\ a & b-a & c-a \\ a^2 & b^2-a^2 & c^2-a^2 \end{vmatrix} \quad \begin{matrix} C_2 = C_2 - C_1 \\ C_3 = C_3 - C_1 \end{matrix}$$

$$= 1\begin{vmatrix} b-a & c-a \\ b^2-a^2 & c^2-a^2 \end{vmatrix}$$

$$= (b-a)(c-a)\begin{vmatrix} 1 & 1 \\ b+a & c+a \end{vmatrix}$$

By placing the common factors $(b-a)$ and $(c-a)$ outside the determinant

$$= (b-a)(c-a)(c+a-b-a)$$
$$= (b-a)(c-a)(c-b)$$
$$= (b-a)(c-a)(a-b)$$

Proved.

Example 2.8. *Using properties of determinants, prove that*

$$\begin{vmatrix} a-b-c & 2a & 2a \\ 2b & b-c-a & 2b \\ 2c & 2c & c-a-b \end{vmatrix} = (a+b+c)^3$$

Solution:

$$\Delta = \begin{vmatrix} a-b-c & 2a & 2a \\ 2b & b-c-a & 2b \\ 2c & 2c & c-a-b \end{vmatrix}$$

$R_1 \to R_1 + R_2 + R_3$

$$= \begin{vmatrix} a+b+c & a+b+c & a+b+c \\ 2b & b-c-a & 2b \\ 2c & 2c & c-a-b \end{vmatrix}$$

$$\therefore \quad \Delta = (a+b+c)\begin{vmatrix} 1 & 1 & 1 \\ 2b & b-c & 2b \\ 2c & 2c & c-a-b \end{vmatrix}$$

$C_2 \to (C_2 - C_1);\ C_3 \to (C_3 - C_1)$

$$= (a+b+c)\begin{vmatrix} 1 & 0 & 0 \\ 2b & -(b+c+a) & 0 \\ 2c & 0 & -(a+b+c) \end{vmatrix}$$

$$= (a+b+c)(b+c+a)(a+b+c) = (a+b+c)^3$$ **Proved.**

2.5 SOLUTION OF LINEAR EQUATIONS BY DETERMINANTS

The properties of determinants are also usefully employed in solving simultaneous linear equations as explained under:

Consider the following equations:

$$a_1x + b_1y = c_1 \quad \text{...(i)}$$
$$a_2x + b_2y = c_2 \quad \text{...(ii)}$$

Multiplying equation (*i*) by b_2 and equation (*ii*) by b_1, we get

$$a_1b_2x + b_1b_2y = c_1b_2$$
$$a_2b_1x + b_1b_2y = b_1c_2$$

or $$x(a_1b_2 - a_2b_1) = c_1b - b_1c_2$$

or $$x = \frac{b_2c_1 - b_1c_2}{a_1b_2 - a_2b_1}$$

Substituting the value of x in equation (*i*), we get

$$a_1\left(\frac{c_1b_2 - b_1c_2}{a_1b_2 - a_2b_1}\right) = b_1y = c_1$$

or $a_1(c_1b_2 - b_1c_2) + b_1y(a_1b_2 - a_2b_1) = c_1(a_1b_2 - a_2b_1)$

or $$y = \frac{c_1(a_1b_2 - a_2b_1) - a_1(c_1b_2 - b_1c_2)}{b_1(a_1b_2 - a_2b_1)}$$

$$= \frac{c_1a_1b_2 - c_1a_2b_1 - a_1c_1b_2 + a_1b_1c_2}{a_1b_1b_1 - a_2b_1b_1}$$

$$= \frac{b_1(a_1c_2 - a_2c_1)}{b_1(a_1b_2 - a_2b_1)}$$

$$\therefore \quad y = \frac{a_1c_2 - a_2c_1}{a_1b_2 - a_2b_1}$$

$$x = \frac{b_2c_1 - b_1c_2}{a_1b_2 - a_2b_1} \text{ and } y = \frac{a_1c_2 - a_2c_1}{a_1b_2 - a_2b_1}$$

In the determinant form, the solution may be expressed as:

$$x = \frac{\begin{vmatrix} c_1 & b_1 \\ c_2 & b_2 \end{vmatrix}}{\begin{vmatrix} a_1 & b_1 \\ a_1 & b_2 \end{vmatrix}} \quad y = \frac{\begin{vmatrix} c_1 & b_1 \\ c_2 & b_2 \end{vmatrix}}{\begin{vmatrix} a_1 & b_1 \\ a_1 & b_2 \end{vmatrix}}$$

We notice that the determinant of the denominator may be obtained from the coefficients and the determinants of the numerators may be obtained by replacing the coefficients of x and y respectively by the constant terms.

The method of solving a system of two equations can be used in exactly the same way to solve a system of 3 equations, 4 equations, ..., n equations.

Let us consider the following three linear equations

$$a_1x + b_1y + c_1z = d_1$$
$$a_2x + b_2y + c_2z = d_2$$
$$a_3x + b_3y + c_3z = d_3$$

then $$x = \frac{\begin{vmatrix} d_1 & b_1 & c_1 \\ d_2 & b_2 & c_2 \\ d_3 & b_3 & c_3 \end{vmatrix}}{\begin{vmatrix} a_1 & b_1 & c_1 \\ a_2 & b_2 & c_2 \\ a_3 & b_3 & c_3 \end{vmatrix}}; \; y = \frac{\begin{vmatrix} a_1 & d_1 & c_1 \\ a_2 & d_2 & c_2 \\ a_3 & d_3 & c_3 \end{vmatrix}}{\begin{vmatrix} a_1 & b_1 & c_1 \\ a_2 & b_2 & c_2 \\ a_3 & b_3 & c_3 \end{vmatrix}}; \; z = \frac{\begin{vmatrix} a_1 & b_1 & d_1 \\ a_2 & b_2 & d_2 \\ a_3 & b_3 & d_3 \end{vmatrix}}{\begin{vmatrix} a_1 & b_1 & c_1 \\ a_2 & b_2 & c_2 \\ a_3 & b_3 & c_3 \end{vmatrix}}$$

This is a Cramer's rule.

$\therefore \; \frac{\Delta_1}{\Delta}; \; y = \frac{\Delta_2}{\Delta}$ and $z = \frac{\Delta_3}{\Delta}$ where Δ_2, Δ_3 are obtained from Δ by substituting the column (d_1, d_2, d_3) for the second and third column of equation (*i*) and (*ii*) is as made columns respectively,

Note: The following points may be noted:

(*i*) While evaluating the value of x, substitute the coefficients of column of constant terms for the column of x terms in the equation.

(*ii*) While evaluating the value of y, substitute the column of constant terms for the column of coefficients of y terms in the equation.... and so on.

(*iii*) The Cramer's rule fails if the determinant of the denominator is zero.

Example. 2.9. *Solve the following simultaneous linear equations by applying cramer's rule.*

$$x + y + z = 4$$
$$x + 2y + z = 6$$
$$x + y + 2z = 5$$

Solution: The solution of the equations is obtained as under

$$x = \frac{\Delta_1}{\Delta}; y = \frac{\Delta_2}{\Delta}; z = \frac{\Delta_3}{\Delta}$$

Here

$$\Delta_1 = \begin{vmatrix} 1 & 1 & 1 \\ 1 & 2 & 1 \\ 1 & 1 & 2 \end{vmatrix} \begin{matrix} = 1(4-1) - 1(2-1) + 1(1-2) \\ = 3 - 1 - 1 = 1 \end{matrix}$$

$$\Delta_2 = \begin{vmatrix} 4 & 1 & 1 \\ 6 & 2 & 1 \\ 5 & 1 & 2 \end{vmatrix} \begin{matrix} = 4(4-1) - 1(12-5) + 1(6-10) \\ = 12 - 7 - 4 = 1 \end{matrix}$$

$$\Delta_3 = \begin{vmatrix} 1 & 4 & 1 \\ 1 & 5 & 1 \\ 1 & 6 & 2 \end{vmatrix} \begin{matrix} = 1(12-5) - 4(2-1) + 1(5-6) \\ = 7 - 4 - 1 = 2 \end{matrix}$$

$$\therefore \quad x = \frac{\Delta_1}{\Delta} = \frac{1}{1} = 1$$

$$y = \frac{\Delta_2}{\Delta} = \frac{2}{1} = 2$$

$$z = \frac{\Delta_3}{\Delta} = \frac{1}{1} = 1$$

$$\therefore \quad \left.\begin{matrix} x = 1 \\ y = 2 \\ z = 1 \end{matrix}\right\} \quad \textbf{Ans.}$$

Example. 2.10. *Solve the following simultaneous linear equations:*

$$x + y + z + w = 0$$
$$x - 2y + 2z + 2w = -6$$
$$2x + y - 2z + 2w = -8$$
$$3x - y + 3z - 3w = 1$$

Solution: The solution of the equations is obtained as under

$$x = \frac{\Delta_1}{\Delta};\ y = \frac{\Delta_2}{\Delta};\ z = \frac{\Delta_3}{\Delta};\ w = \frac{\Delta_4}{\Delta}$$

Here
$$\Delta = \begin{vmatrix} 1 & 1 & 1 & 1 \\ 1 & -2 & 2 & 2 \\ 2 & 1 & -2 & 2 \\ 3 & -1 & 3 & -3 \end{vmatrix}$$

$$= \begin{vmatrix} 1 & 1 & 0 & 0 \\ 1 & -3 & 1 & 1 \\ 2 & -1 & -4 & 0 \\ 3 & -4 & 0 & -6 \end{vmatrix} \begin{matrix} c_2 = c_2 - c_1\ ;\ c_3 - c_1 \\ c_4 = c_4 - c_1 \end{matrix}$$

$$= \begin{vmatrix} -3 & 1 & 1 \\ -1 & -4 & 0 \\ -4 & 0 & -6 \end{vmatrix} \begin{matrix} = -3(24-0) - 1(6-0) + 1(0-16) \\ = -72 - 6 - 16 = -94 \end{matrix}$$

$$\Delta_1 = \begin{vmatrix} 0 & 1 & 1 & 1 \\ -6 & -2 & 2 & 2 \\ -8 & 1 & -2 & 2 \\ 1 & -1 & 3 & 3 \end{vmatrix}$$

$$= \begin{vmatrix} 0 & 1 & 1 & 0 \\ -6 & -2 & 2 & 4 \\ -8 & 1 & -3 & 1 \\ 1 & -1 & 4 & -2 \end{vmatrix} c_3 = c_3 - c_2\ ;\ c_4 - c_2$$

$$= \begin{vmatrix} 0 & 1 & 0 & 0 \\ -6 & -2 & 4 & 4 \\ -8 & 1 & -3 & 1 \\ 1 & -1 & 4 & -2 \end{vmatrix}$$

$$= \begin{vmatrix} -6 & 4 & 4 \\ -8 & -3 & 1 \\ 1 & 4 & -2 \end{vmatrix} \begin{matrix} = -6(6-4) - 4(16-1) + 4(-32+3) \\ = -12 - 60 - 116 = 12 + 60 + 116 = 188 \end{matrix}$$

$$\therefore \qquad x = \frac{\Delta_1}{\Delta} = \frac{186}{-94} = -2$$

$$\Delta_2 = \begin{vmatrix} 1 & 0 & 1 & 1 \\ 1 & -6 & 2 & 2 \\ 2 & -8 & -2 & 2 \\ 3 & 1 & 3 & -3 \end{vmatrix}$$

$$= \begin{vmatrix} 1 & 0 & 0 & 0 \\ 1 & -6 & 1 & 1 \\ 2 & -8 & -4 & 0 \\ 3 & 1 & 0 & -6 \end{vmatrix} c_3 = c_3 - 1 ; c_4 = c_4 - c_1$$

$$= \begin{vmatrix} -6 & 1 & 1 \\ -8 & -4 & 0 \\ 1 & 0 & -6 \end{vmatrix} \begin{matrix} = -6(24-0) - 1(48-0) + 1(0+4) \\ = -144 - 48 + 4 = -188 \end{matrix}$$

$$\therefore \qquad y = \frac{\Delta_2}{\Delta} = \frac{186}{-94} = 2$$

$$\Delta_2 = \begin{vmatrix} 1 & 0 & 0 & 1 \\ 1 & -2 & -6 & 2 \\ 2 & 1 & -8 & 2 \\ 3 & -1 & 0 & -3 \end{vmatrix} = \begin{vmatrix} 1 & 0 & 0 & 0 \\ 1 & -3 & -6 & 1 \\ 2 & -1 & -8 & 0 \\ 3 & -4 & 1 & -6 \end{vmatrix} c_2 = c_2 - c_1 ; c_4 - c_1$$

$$= \begin{vmatrix} -3 & -6 & 1 \\ -1 & -8 & 0 \\ -4 & 1 & -6 \end{vmatrix} \begin{matrix} = -3(48-0) + 6(6-0) + 1(-1-32) \\ = -144 + 36 \ 33 = -141 \end{matrix}$$

$$\therefore \qquad x = \frac{\Delta_3}{\Delta} = \frac{-141}{-94} = -2$$

$$\Delta_2 = \begin{vmatrix} 1 & 1 & 1 & 0 \\ 1 & -2 & 2 & -6 \\ 2 & 1 & -2 & -8 \\ 3 & 1 & 3 & 1 \end{vmatrix}$$

$$= \begin{vmatrix} 1 & 0 & 0 & 0 \\ 1 & -3 & 1 & -6 \\ 2 & -1 & -4 & -8 \\ 3 & -4 & 0 & 1 \end{vmatrix} c_2 = c_2 - c_1 \,;\, c_3 = c_3 - c_1$$

$$= \begin{vmatrix} -3 & 1 & -6 \\ -1 & -4 & -8 \\ -4 & 0 & 1 \end{vmatrix} \begin{aligned} &= -3(-4+0) - 1(-1-32) - 6(0-16) \\ &= -12 + 33 + 96 = 144 \end{aligned}$$

$$\therefore \qquad y = \frac{\Delta_4}{\Delta} = \frac{141}{-94} = -1.5$$

$$\left.\begin{aligned} x &= -2 \\ y &= 2 \\ z &= 1.5 \\ w &= -1.5 \end{aligned}\right\} \textbf{Ans.}$$

❑❑❑

3 Matrices

3.1 INTRODUCTION

The system of numbers arranged in a rectangular array in rows and columns and enclosed in a pair of parentheses is called a matrix. For an example,

$$A = \begin{array}{c} \\ \end{array}\begin{matrix} \text{col.1} & \text{col.2} & \text{col.3} & \text{col.4} \end{matrix}$$

$$A = \begin{bmatrix} 2 & 3 & 4 & 5 \\ 3 & 4 & 5 & 1 \\ 4 & 1 & 3 & 7 \\ 5 & 2 & 6 & 3 \end{bmatrix} \begin{matrix} \text{......... row 1} \\ \text{......... row 2} \\ \text{......... row 3} \\ \text{......... row 4} \end{matrix}$$

This matrix has 4 rows and 4 columns and it contains 4 × 4 = 16 elements. It is called a 4 × 4 matrix. If a matrix has 2 rows and three column it is known as 2 × 3 matrix to be read as 2 by 3 matrix. Similarly, if the matrix has 4 rows and five columns, it is known as 4 × 5 matrix or 4 by 5 matrix. In case the elements of the matrix are described by double subscripts, the first subscript determines the row and the second subscript determines the column in which the particular element lies. For example, the element a_{ij} lies in the ith row and jth column.

3.2 TYPES OF MATRICES

The matrices are of the following types:

(1) **Row matrix.** The matrix which has only one row and any number of columns, is called a *row matrix, i.e.,*

[3 7 5 2], [4 5 6 9], [1 2 3 4], etc.

(2) **Column matrix.** The matrix which has only one column and any number of rows, is called a *column matrix, i.e.,*

$$\begin{bmatrix} 3 \\ 7 \\ 5 \end{bmatrix}, \begin{bmatrix} 2 \\ 6 \\ 7 \\ 5 \end{bmatrix}, \begin{bmatrix} 3 \\ 7 \\ 8 \\ 6 \end{bmatrix}, \text{ etc.}$$

(3) **Zero matrix (or null matrix).** The matrix of which every element is zero, is called a *zero matrix* or *null matrix., i.e.,*

$$\begin{bmatrix} 0 & 0 & 0 & 0 \\ 0 & 0 & 0 & 0 \\ 0 & 0 & 0 & 0 \end{bmatrix}$$

(4) **Square matrix.** The matrix in which the number of rows is equal to the number of columns, is called a *square matrix, i.e.,*

$$\begin{bmatrix} 4 & 3 & 7 & 8 \\ 2 & 1 & 6 & 5 \\ 3 & 7 & 1 & 6 \\ 2 & 5 & 7 & 4 \end{bmatrix}$$

(5) **Diagonal matrix.** The matrix whose non-diagonal elements are all zeros, is called a *diagonal matrix, i.e.,*

$$\begin{bmatrix} 5 & 0 & 0 & 0 \\ 0 & 7 & 0 & 0 \\ 0 & 0 & 6 & 0 \\ 0 & 0 & 0 & 2 \end{bmatrix}$$

(6) **Unit matrix.** The square matrix in which all the diagonal elements are unity and the non-diagonal elements are zeros, is called a *unit matrix* or *identity matrix, i.e.,*

$$\begin{bmatrix} 1 & 0 & 0 \\ 0 & 1 & 0 \\ 0 & 0 & 1 \end{bmatrix}, \begin{bmatrix} 1 & 0 & 0 & 0 \\ 0 & 1 & 0 & 0 \\ 0 & 0 & 1 & 0 \\ 0 & 0 & 0 & 1 \end{bmatrix}, \begin{bmatrix} 1 & 0 \\ 0 & 1 \end{bmatrix}$$

(7) **Symmetric matrix.** The square matrix in which, elements of rows are also elements of columns, is called *a symmetric matrix, i.e.,*

$$\begin{bmatrix} a & h & g \\ h & b & f \\ g & f & c \end{bmatrix}, \begin{bmatrix} 5 & 3 & 7 & 2 \\ 3 & 4 & 3 & 8 \\ 7 & 3 & 7 & 2 \\ 2 & 8 & 2 & 1 \end{bmatrix}, \text{etc.}$$

(8) **Skew symmetric matrix.** The square matrix in which diagonal elements are each zero and the elements in the upper triangular portion are respectively equal to lower triangular portion with an opposite sign, is called a *skew symmetric matrix., i.e.,*

$$\begin{bmatrix} 0 & -8 & -3 & -1 \\ 8 & 0 & -2 & -7 \\ 3 & 2 & 0 & -5 \\ 1 & 7 & 5 & 0 \end{bmatrix}, \begin{bmatrix} 0 & -c & -d \\ c & 0 & -e \\ d & e & 0 \end{bmatrix}.$$

(9) **Triangular matrix.** The square matrix whose elements above or below the leading diagonal are zeroes, is called a *triangular matrix.* If the zero elements are below the leading diagonal, it is called an *upper triangular matrix* and the one with zero elements in upper side of the leading diagonal, is called *lower triangular matrix, i.e.,*

$$\begin{bmatrix} 2 & 3 & 4 \\ 0 & 5 & 8 \\ 0 & 0 & 7 \end{bmatrix} \qquad \begin{bmatrix} 4 & 0 & 0 & 0 \\ 0 & 5 & 0 & 0 \\ 2 & 3 & 7 & 0 \\ 8 & 7 & 1 & 6 \end{bmatrix}$$

Upper matrix. Lower Triangular matrix.

(10) **Transpose of a matrix.** The matrix which is obtained by changing the rows into columns and columns into rows of a matrix, is called the *transpose of the matrix.* It is denoted by A′, *i.e.,*

$$A = \begin{bmatrix} 7 & 3 & 2 & 5 \\ 1 & 4 & 6 & 8 \\ 3 & 7 & 2 & 7 \end{bmatrix} \qquad A' = \begin{bmatrix} 7 & 1 & 3 \\ 3 & 4 & 7 \\ 2 & 6 & 2 \\ 5 & 8 & 7 \end{bmatrix}$$

(11) **Orthogonal matrix.** The square matrix is said to be orthogonal matrix if the product of the matrix and its transpose is an identity matrix, *i.e.,*

$$[A] \cdot [A'] = 1$$

(12) **Order of the matrix.** If the determinant of a matrix consists of one row and one column, it is said to be of the first order. If the determinant of the matrix consists of two rows and two columns, it is said to be of the second order.

If the determinant of the matrix consists of three rows and three columns, it is said to be the third order.

(13) **Co-factor of an element of a matrix.** The minor of an element is the determinant that is left after removing the row and the column containing the particular element.

Co-factor of an element of a matrix is the minor of that element but with proper sign.

The sign of the co-factor may be decided as under:

Add up the number of the row and the column in which the element occurs. If the sum is even, the sign of the co-factor of that element is positive. If the sum is odd, the sign of the co-factor is negative.

For example,

$$\text{Let} \qquad A = \begin{bmatrix} a & b & c \\ d & e & f \\ g & h & i \end{bmatrix}$$

$$\text{then co-factor of } \quad d = \begin{bmatrix} b & c \\ h & i \end{bmatrix}$$

The element d occurs in the second row and the first column.

∴ Sum of the number of row and column = 2 + 1 = 3 an odd.

$$\therefore \qquad \text{Co-factor of } d = \begin{bmatrix} b & c \\ h & i \end{bmatrix}$$

(14) **Non-singular matrix.** A square matrix is said to be non-singular if $|A| \neq 0$. Otherwise A is said to be singular.

3.3 ADDITION OF MATRICES

Two matrices A, B are conformable for addition when each has the same number of rows and each has the same number of columns.

The sum of two matrices A and B may be obtained by adding the corresponding elements of matrices A and B provided the matrices are of the same order, *i.e.*,

$$\text{Let} \qquad A = \begin{bmatrix} 3 & 7 & 5 \\ 1 & 3 & 4 \\ 2 & 6 & 3 \end{bmatrix} \text{ and } B = \begin{bmatrix} 8 & 1 & 2 \\ 3 & 7 & 6 \\ 5 & 4 & 3 \end{bmatrix}$$

$$\therefore \qquad A + B = \begin{bmatrix} 3+8 & 7+1 & 5+2 \\ 1+3 & 3+7 & 4+6 \\ 2+5 & 6+4 & 3+3 \end{bmatrix} = \begin{bmatrix} 11 & 8 & 7 \\ 4 & 10 & 10 \\ 7 & 10 & 6 \end{bmatrix} \quad \textbf{Ans.}$$

3.4 SUBTRACTION OF MATRICES

Two matrices A, B are conformable for subtraction when each has the same number of rows and each has the same number of columns.

The matrix B may be subtracted from the matrix A by taking the difference of corresponding elements of matrix A and matrix B, provided the two matrices are of the same order, *i.e.*,

Let

$$A = \begin{bmatrix} 3 & 7 & 5 & 2 \\ 1 & 3 & 4 & 6 \\ 3 & 2 & 1 & 2 \end{bmatrix} \text{ and } B = \begin{bmatrix} 1 & 3 & 5 & 3 \\ 2 & 6 & 4 & 2 \\ 6 & 3 & 7 & 4 \end{bmatrix}$$

$$\therefore \quad A - B = \begin{bmatrix} 3-1 & 7-3 & 5-5 & 2-3 \\ 1-2 & 3-6 & 4-4 & 6-2 \\ 3-6 & 2-3 & 1-7 & 2-4 \end{bmatrix}$$

$$= \begin{bmatrix} 2 & 4 & 0 & -1 \\ -1 & -3 & 0 & 4 \\ -3 & -1 & -6 & -2 \end{bmatrix}$$

Note: The following points may be noted:

(*i*) A matrix – A is that metric whose elements, are those of A, each multiplied by – 1.

(*ii*) Matrix A – B is defined as matrix A + (– B).

3.5 COMMUTATIVE, ASSOCIATIVE LAWS FOR ADDITION AND DISTRIBUTIVE LAWS IN MATRICES

Let A, B, C be three matrices conformable for addition then

(*i*) A + B = B + A

(*ii*) (A + B) = C = A + (B + C)

(*iiii*) $m(A + B) = mA + mB$.

Let $$A = \begin{bmatrix} a_{11} & a_{12} & a_{13} \\ a_{21} & a_{22} & a_{23} \end{bmatrix}, B = \begin{bmatrix} b_{11} & b_{12} & b_{13} \\ b_{21} & b_{22} & b_{23} \end{bmatrix}$$

and $$C = \begin{bmatrix} c_{11} & c_{12} & c_{13} \\ c_{21} & c_{22} & c_{23} \end{bmatrix}$$

(*i*) $$A + B = \begin{bmatrix} a_{11}+b_{11} & a_{12}+b_{12} & a_{13}+b_{13} \\ a_{21}+b_{21} & a_{22}+b_{22} & a_{23}+b_{23} \end{bmatrix}$$

$$= \begin{bmatrix} b_{11}+a_{11} & b_{12}+a_{12} & b_{13}+a_{13} \\ b_{21}+a_{21} & b_{22}+a_{22} & b_{23}+a_{23} \end{bmatrix}$$

$$= B + A$$

(*ii*) (A + B) + C

$$= \begin{bmatrix} (a_{11}+b_{11})+c_{11} & (a_{12}+b_{12})+c_{12} & (a_{13}+b_{13})+c_{13} \\ (a_{21}+b_{21})+c_{21} & (a_{22}+b_{22})+c_{22} & (a_{23}+b_{23})+c_{23} \end{bmatrix}$$

$$= \begin{bmatrix} a_{11} + (b_{11} + c_{11}) & a_{12} + (b_{12} + c_{12}) & a_{13} + (b_{13} + c_{13}) \\ a_{21} + (b_{21} + c_{21}) & a_{22} + (b_{22} + c_{22}) & a_{23} + (b_{23} + c_{23}) \end{bmatrix}$$

$$= \text{A} + (\text{B} + \text{C})$$

(*iii*) $$m\text{A} + m\text{B} = \begin{bmatrix} ma_{11} + mb_{11} & ma_{12} + mb_{12} & ma_{13} + mb_{13} \\ ma_{21} + mb_{21} & ma_{22} + mb_{22} & ma_{23} + mb_{23} \end{bmatrix}$$

$$= m(\text{A} + \text{B})$$

3.6 SCALAR MULTIPLE OF A MATRIX

A matrix may be multiplied by a scalar quantity by multiplying each elements of the given matrix by the scalar quantity, *i.e.*,

Let $$\text{A} = \begin{bmatrix} 5 & 3 & 7 & 2 \\ 1 & 4 & 3 & 5 \\ 3 & 2 & 6 & 7 \end{bmatrix} \text{ then}$$

$$3\text{ A} = 3\begin{bmatrix} 5 & 3 & 7 & 2 \\ 1 & 4 & 3 & 5 \\ 3 & 2 & 6 & 7 \end{bmatrix} = \begin{bmatrix} 3\times5 & 3\times3 & 3\times7 & 3\times2 \\ 3\times1 & 3\times4 & 3\times3 & 3\times5 \\ 3\times3 & 3\times2 & 3\times6 & 3\times7 \end{bmatrix}$$

$$= \begin{bmatrix} 15 & 9 & 21 & 6 \\ 3 & 12 & 9 & 15 \\ 9 & 6 & 18 & 21 \end{bmatrix}$$

Example 3.1. *Find the matrix A + B when*

$$A = \begin{bmatrix} 2 & 3 & 4 \\ 1 & 2 & 3 \\ 4 & 1 & 2 \end{bmatrix} \text{ and } B = \begin{bmatrix} 2 & 3 & -1 \\ 4 & -2 & 3 \\ -2 & 3 & 5 \end{bmatrix}$$

Solution: $$\text{A} = \begin{bmatrix} 2 & 3 & 4 \\ 1 & 2 & 3 \\ 4 & 1 & 2 \end{bmatrix} \text{ and B} = \begin{bmatrix} 2 & 3 & -1 \\ 4 & -2 & 3 \\ -2 & 3 & 5 \end{bmatrix}$$

$\therefore$ $$\text{A} + \text{B} = \begin{bmatrix} 2+2 & 3+3 & 4-1 \\ 1+4 & 2-2 & 3+3 \\ 4-2 & 1+3 & 2+5 \end{bmatrix}$$

$$= \begin{bmatrix} 4 & 6 & 3 \\ 5 & 0 & 6 \\ 2 & 4 & 7 \end{bmatrix}$$ **Ans.**

Example 3.2. *Find the matrix A + B when*

$$A = \begin{bmatrix} 1 & 1 & 1 \\ 1 & 2 & 3 \\ 1 & 4 & 9 \end{bmatrix} \text{ and } B = \begin{bmatrix} 2 & 5 & 3 \\ 3 & 1 & 2 \\ 1 & 2 & 1 \end{bmatrix}$$

Solution:

$$A = \begin{bmatrix} 1 & 1 & 1 \\ 1 & 2 & 3 \\ 1 & 4 & 9 \end{bmatrix} \text{ and } B = \begin{bmatrix} 2 & 5 & 3 \\ 3 & 1 & 2 \\ 1 & 2 & 1 \end{bmatrix}$$

$$\therefore \quad A + B = \begin{bmatrix} 1+2 & 1+5 & 1+3 \\ 1+3 & 2+1 & 3+2 \\ 1+1 & 4+2 & 9+1 \end{bmatrix}$$

$$= \begin{bmatrix} 3 & 6 & 4 \\ 4 & 3 & 5 \\ 2 & 6 & 10 \end{bmatrix} \quad \textbf{Ans.}$$

Example 3.3. *Find the matrix A – B when*

$$A = \begin{bmatrix} 4 & 6 & 3 \\ 3 & 5 & 4 \\ 3 & 0 & 1 \end{bmatrix} \text{ and } B = \begin{bmatrix} 2 & 3 & 4 \\ 1 & 2 & 3 \\ 0 & 1 & 5 \end{bmatrix}$$

Solution:

$$A = \begin{bmatrix} 4 & 6 & 3 \\ 3 & 5 & 4 \\ 3 & 0 & 1 \end{bmatrix} \text{ and } B = \begin{bmatrix} 2 & 3 & 4 \\ 1 & 2 & 3 \\ 0 & 1 & 5 \end{bmatrix}$$

$$\therefore \quad A - B = \begin{bmatrix} 4-2 & 6-3 & 3-4 \\ 3-1 & 5-2 & 4-3 \\ 3-0 & 0-1 & 1-5 \end{bmatrix}$$

$$= \begin{bmatrix} 2 & 3 & -1 \\ 2 & 3 & 1 \\ 3 & -1 & -4 \end{bmatrix} \quad \textbf{Ans.}$$

3.7 MULTIPLICATION OF MATRICES

The matrix A may be multiplied by another matrix B only if *the number of columns in A is equal to the number of rows in B.*

To get the elements of the first row of the resultant matrix AB, proceed as under:

(*i*) Multiply each element of the first row of matrix A by the corresponding elements of the first column of matrix B and add all the terms so obtained, to get the first element of the first row of the resultant matrix AB.

(*ii*) Multiply each element of the first row of matrix A by the corresponding elements of the second column of matrix B and add all the terms so obtained, to get the second element of the first row of the resultant matrix AB.

(*iiii*) Similarly, multiply each element of the first row of matrix A by the corresponding elements of *nth* column of matrix B and add all the terms so obtained, to get the *nth* element of the first row of the resultant matrix AB.

To get the elements of the second row of the resultant matrix AB, proceed as under:

(*i*) Multiply each element of the second row of matrix A by the corresponding elements of the first column of the matrix B and add all the terms so obtained, to get the first element of the second row of the resultant matrix AB.

(*ii*) Multiply each element of the second row of matrix A by the corresponding elements of the second column of matrix B and add all the terms so obtained, to get the second element of the second row of the resultant matrix AB.

(*iiii*) Similarly, multiply each element of the second row of matrix A by the corresponding elements of *nth* column of matrix B and add all the terms so obtained, to get the *nth* element of the second row of the resultant matrix AB.

The procedure may be continued to obtain the values of elements of further rows of the matrix AB, if any.

In the product AB, the matrix B is said to be *premultiplied* by A and A is said to be *post-multiplied* by B. In multiplication of matrices, it is not necessary that AB = BA.

To understand the procedure of multiplication, the method of the solution of the following examples may be followed.

Example 3.4. *Find the product AB and BA, where*

$$(i)\ A = \begin{bmatrix} 0 & 2 & 1 \\ 3 & 2 & 1 \\ 4 & 3 & 2 \end{bmatrix} \quad B = \begin{bmatrix} 1 & -2 \\ 0 & -1 \\ -1 & 2 \end{bmatrix}$$

$$(ii)\ A = \begin{bmatrix} -2 & 1 & 3 \\ 2 & -4 & 5 \end{bmatrix} \quad B = \begin{bmatrix} 3 & 1 \\ 0 & -1 \\ 2 & 4 \end{bmatrix}$$

(iiii) $A = \begin{bmatrix} 1 & -2 & 3 \\ 2 & 3 & -1 \\ -3 & 1 & 2 \end{bmatrix}$ $B = \begin{bmatrix} 1 & 0 & 2 \\ 0 & 1 & 2 \\ 1 & 2 & 0 \end{bmatrix}$

Solution:

(i) Matrix A Matrix B

$$\text{Product AB} = \begin{bmatrix} 0 & 2 & 1 \\ 3 & 2 & 1 \\ 4 & 3 & 2 \end{bmatrix} \times \begin{bmatrix} 1 & -2 \\ 0 & -1 \\ -1 & 2 \end{bmatrix}$$

$$\text{AB} = \begin{bmatrix} 0+0-1 & 0-2+2 \\ 3+0-1 & -6-2+2 \\ 4+0-2 & -8-3+4 \end{bmatrix} = \begin{bmatrix} -1 & 0 \\ 2 & -6 \\ 2 & -7 \end{bmatrix} \quad \textbf{Ans.}$$

As the number of columns in **B** is 2 and the number of rows in **A** is 3, the product BA is not possible.

(ii) Matrix A Matrix B

$$\text{Product AB} = \begin{bmatrix} -2 & 1 & 3 \\ 2 & 4 & 5 \end{bmatrix} \times \begin{bmatrix} 3 & 1 \\ 0 & -1 \\ 2 & 4 \end{bmatrix}$$

$$\text{AB} = \begin{bmatrix} -6+0+6 & -2+-1+12 \\ 6+0+10 & 2+4+20 \end{bmatrix} = \begin{bmatrix} 0 & 9 \\ 16 & 26 \end{bmatrix} \quad \textbf{Ans.}$$

$$\text{BA} = \begin{bmatrix} -6+2 & 3-4 & 9+5 \\ 0-2 & 0+4 & 0-5 \\ -4+8 & 2-16 & 6+20 \end{bmatrix} = \begin{bmatrix} -4 & -1 & 14 \\ -2 & 4 & -5 \\ 4 & -14 & 26 \end{bmatrix} \quad \textbf{Ans.}$$

(iii) Matrix A Matrix B

$$\text{Product AB} = \begin{bmatrix} 1 & -2 & 3 \\ 2 & 3 & -1 \\ -3 & 1 & 2 \end{bmatrix} \times \begin{bmatrix} 1 & 0 & 2 \\ 0 & 1 & 2 \\ 1 & 2 & 0 \end{bmatrix}$$

$$\text{AB} = \begin{bmatrix} 1+0+3 & 0-2+6 & 2-4+0 \\ 2+0+-1 & 0+3-2 & 4+6+0 \\ -3+0+2 & 0+1+4 & -6+2+0 \end{bmatrix} \quad \textbf{Ans.}$$

$$= \begin{bmatrix} 4 & 4 & -2 \\ 1 & 1 & 10 \\ -1 & 5 & -4 \end{bmatrix} \quad \textbf{Ans.}$$

$$BA = \begin{bmatrix} 1+0+-6 & -2+0+2 & 3+0+4 \\ 0+2-6 & 0+3+2 & 0-1+4 \\ 1+4+0 & -2+6+0 & 3-2+0 \end{bmatrix}$$

$$= \begin{bmatrix} -5 & 0 & 7 \\ -4 & 5 & 3 \\ 5 & 4 & 1 \end{bmatrix} \quad \textbf{Ans.}$$

Example 3.5. *Find the product of matrices A and B as under:*

$$A = \begin{bmatrix} 0 & 5 & 3 \\ 2 & 1 & 4 \\ 1 & 6 & 2 \end{bmatrix} \text{ and } B = \begin{bmatrix} 7 & 3 & 2 \\ 1 & 5 & 6 \\ 3 & 2 & 1 \end{bmatrix}$$

Solution:

$$AB = \begin{bmatrix} 0 & 5 & 3 \\ 2 & 1 & 4 \\ 1 & 6 & 2 \end{bmatrix} \times \begin{bmatrix} 7 & 3 & 2 \\ 1 & 5 & 6 \\ 3 & 2 & 1 \end{bmatrix}$$

$$= \begin{bmatrix} 0\times7+5\times1+3\times3 & 0\times3+5\times5+3\times2 & 0\times2+5\times6+3\times1 \\ 2\times7+1\times1+4\times3 & 2\times3+1\times5+4\times2 & 2\times2+1\times6+4\times1 \\ 1\times7+6\times1+2\times3 & 1\times3+6\times5+2\times2 & 1\times2+6\times6+2\times1 \end{bmatrix}$$

$$= \begin{bmatrix} 0+5+9 & 0+25+6 & 0+30+3 \\ 14+1+12 & 6+5+8 & 4+6+4 \\ 7+6+6 & 3+30+4 & 2+36+2 \end{bmatrix}$$

$$= \begin{bmatrix} 14 & 31 & 33 \\ 27 & 19 & 14 \\ 19 & 37 & 40 \end{bmatrix} \quad \textbf{Ans.}$$

Example 3.6. *Show that the following matrices are orthogonal.*

(i) $\begin{bmatrix} \cos\alpha & \sin\alpha \\ \sin\alpha & \cos\alpha \end{bmatrix}$, *(ii)* $\frac{1}{3}\begin{bmatrix} 1 & 2 & 2 \\ 2 & 1 & -2 \\ 2 & -2 & +1 \end{bmatrix}$ *and (iii)* $\frac{1}{3}\begin{bmatrix} -2 & 1 & 2 \\ 2 & 2 & 1 \\ 1 & -2 & 2 \end{bmatrix}$

Solution:

(i) $$A = \begin{bmatrix} \cos\alpha & \sin\alpha \\ -\sin\alpha & \cos\alpha \end{bmatrix} \text{and } A = \begin{bmatrix} \cos\alpha & -\sin\alpha \\ \sin\alpha & -\cos\alpha \end{bmatrix}$$

$$AA' = \begin{bmatrix} \cos\alpha & \sin\alpha \\ -\sin\alpha & \cos\alpha \end{bmatrix} \times \begin{bmatrix} \cos\alpha & -\sin\alpha \\ \sin\alpha & -\cos\alpha \end{bmatrix}$$

$$= \begin{bmatrix} \cos\alpha\cdot\cos\alpha + \sin\alpha\cdot\sin\alpha, & \cos\alpha(-\sin\alpha) + \sin\alpha\cos\alpha \\ -\sin\alpha\cdot\cos\alpha + \cos\alpha\cdot\sin\alpha, & -\sin\alpha(-\sin\alpha) + \cos\alpha\cos\alpha \end{bmatrix}$$

$$= \begin{bmatrix} \cos^2\alpha + \sin^2\alpha & -\sin\alpha\cos\alpha + \sin\alpha\cos\alpha \\ -\sin\alpha\cos\alpha + \sin\alpha\cos\alpha & \sin^2\alpha + \cos^2\alpha \end{bmatrix}$$

$$= \begin{bmatrix} 1 & 0 \\ 0 & 1 \end{bmatrix} = I$$

Since AA′ = I, A is an orthogonal matrix. **Proved.**

(*ii*) $A = \frac{1}{3}\begin{bmatrix} 1 & 2 & 2 \\ 2 & 1 & -2 \\ 2 & -2 & 1 \end{bmatrix} \quad \therefore \quad A' = \frac{1}{3}\begin{bmatrix} 1 & 2 & 2 \\ 2 & 1 & -2 \\ 2 & -2 & 1 \end{bmatrix}$

$$AA' = \frac{1}{3}\begin{bmatrix} 1 & 2 & 2 \\ 2 & 1 & -2 \\ 2 & -2 & 1 \end{bmatrix} \times \frac{1}{3}\begin{bmatrix} 1 & 2 & 2 \\ 2 & 1 & -2 \\ 2 & -2 & 1 \end{bmatrix}$$

$$\frac{1}{9}\begin{bmatrix} 1\times1+2\times2+2\times2, & 1\times2+2\times1-2\times2, & 1\times-2+2\times-2+2\times1 \\ 2\times1+1\times2+(-2)(2), & 2\times2+1\times1+(-2)(-2), & 2\times2-1\times2-2\times1 \\ 2\times1+2(-2)+1+1\times2, & 2\times2-2\times1+1(-2), & 2\times2+2\times2+1\times1 \end{bmatrix}$$

$$= \frac{1}{9}\begin{bmatrix} 9 & 0 & 0 \\ 0 & 9 & 0 \\ 0 & 0 & 9 \end{bmatrix} = \begin{bmatrix} 1 & 0 & 0 \\ 0 & 1 & 0 \\ 0 & 0 & 1 \end{bmatrix} = I$$

Since AA′ = I, A is the orthogonal matrix. **Proved.**

(*iii*) $A = \frac{1}{3}\begin{bmatrix} -2 & 1 & 2 \\ 2 & 2 & 1 \\ 1 & -2 & 2 \end{bmatrix} \quad A' = \frac{1}{3}\begin{bmatrix} -2 & 2 & 1 \\ 1 & 2 & -2 \\ 2 & 1 & 2 \end{bmatrix}$

$$AA' = \frac{1}{3}\begin{bmatrix} -2 & 1 & 2 \\ 2 & 2 & 1 \\ 1 & -2 & 2 \end{bmatrix} \times \frac{1}{3}\begin{bmatrix} -2 & 2 & 1 \\ 1 & 2 & -2 \\ 2 & 1 & 2 \end{bmatrix}$$

$$= \frac{1}{9}\begin{bmatrix} -2 & 1 & 2 \\ 2 & 1 & 1 \\ 1 & -2 & 2 \end{bmatrix} \times \begin{bmatrix} -2 & 2 & 1 \\ 1 & 2 & -2 \\ 2 & 1 & 2 \end{bmatrix}$$

$$\frac{1}{9}\begin{bmatrix} -2\times(-2)+1\times1+2\times2, & (-2)(2)+1\times2+2\times1, & -2\times1+1(-2)+2\times1 \\ 2\times(-2)+2\times1+1\times2, & 2\times2+2\times2+1\times1, & 2\times1+2(-2)+1\times2 \\ 1\times(-2)+(-2)(1)+2\times2, & 1\times2+(-2)2+2\times1, & 1\times1+(-2)\times(-2)+2\times2 \end{bmatrix}$$

$$= \frac{1}{9}\begin{bmatrix} 9 & 0 & 0 \\ 0 & 9 & 0 \\ 0 & 0 & 9 \end{bmatrix} = \begin{bmatrix} 1 & 0 & 0 \\ 0 & 1 & 0 \\ 0 & 0 & 1 \end{bmatrix} = I$$

Since AA′ = I, A is the orthogonal matrix. **Proved.**

3.8 DETERMINANT OF A MATRIX

The determinant of a matrix

$$A = \begin{bmatrix} a_1 & a_2 & a_3 \\ b_1 & b_2 & b_3 \\ c_1 & c_2 & c_3 \end{bmatrix} \text{ is } |A| = \begin{vmatrix} a_1 & a_2 & a_3 \\ b_1 & b_2 & b_3 \\ c_1 & c_2 & c_3 \end{vmatrix}$$

i.e., the determinant of the matrix may be obtained by simply replacing the brackets by two straight lines.

3.9 ADJOINT OF A SQUARE MATRIX

The adjoint of a square matrix

$$A = \begin{bmatrix} a_1 & a_2 & a_3 \\ b_1 & b_2 & b_3 \\ c_1 & c_2 & c_3 \end{bmatrix}$$ may be obtained in the following steps:

(*i*) Obtain the determinant of the given matrix.

(*ii*) Form a matrix of cofactors of the determinant of the matrix.

(*iiii*) The transpose of the matrix of cofactors, is the required adjoint of the matrix.

If the matrix of cofactors of the determinant, is

$$\begin{bmatrix} -7 & 11 & 23 \\ 14 & 13 & -4 \\ 0 & -14 & -7 \end{bmatrix}$$

Then, adjoint of the matrix, *i.e.*, adj. $A = \begin{bmatrix} -7 & 14 & 0 \\ 11 & 13 & -14 \\ 23 & -4 & -7 \end{bmatrix}$

Example 3.7. *Find the adjoint of the following matrix*

$$A = \begin{bmatrix} 2 & 5 & 3 \\ 3 & 1 & 2 \\ 1 & 2 & 1 \end{bmatrix}$$

Solution: The determinant of the matrix

$$A = |A| = \begin{bmatrix} 2 & 5 & 3 \\ 3 & 1 & 2 \\ 1 & 2 & 1 \end{bmatrix}$$

Cofactors of the determinant are:

$$\begin{aligned}
A_1 &= +(1 \times 1 - 2 \times 2) = 3 \\
A_2 &= -(3 \times 1 - 1 \times 2) = -1 \\
A_3 &= +(3 \times 2 - 1 \times 1) = +5 \\
B_1 &= -(5 \times 1 - 2 \times 3) = +1 \\
B_2 &= +(2 \times 1 - 1 \times 3) = -1 \\
B_3 &= -(2 \times 2 - 1 \times 5) = +1 \\
C_1 &= +(5 \times 2 - 1 \times 3) = +7 \\
C_2 &= -(2 \times 2 - 3 \times 3) = +5 \\
C_3 &= +(2 \times 1 - 3 \times 5) = -13
\end{aligned}$$

$\therefore$ Matrix of the cofactors

$$= \begin{bmatrix} -3 & -1 & 5 \\ 1 & -1 & 1 \\ 7 & 5 & -13 \end{bmatrix}$$

$$\therefore \qquad \text{Adj. } A = \begin{bmatrix} -3 & 1 & 7 \\ -1 & -1 & 5 \\ 5 & 1 & -13 \end{bmatrix} \quad \textbf{Ans.}$$

Example 3.8. *Find the adjoint of the matrix*

$$A = \begin{bmatrix} 3 & 5 & 7 \\ 2 & 1 & 6 \\ 4 & 2 & 5 \end{bmatrix}$$

Solution:

(*i*) The determinant of the matrix A

$$|A| = \begin{vmatrix} 3 & 5 & 7 \\ 2 & 1 & 6 \\ 4 & 2 & 5 \end{vmatrix}$$

(*ii*) Cofactors of the determinant are:

$$A_1 = +\begin{vmatrix} 1 & 6 \\ 2 & 5 \end{vmatrix} = (1 \times 5 - 2 \times 6) = -7$$

$$A_2 = +\begin{vmatrix} 2 & 6 \\ 4 & 5 \end{vmatrix} = +(2 \times 5 - 4 \times 6) = -10 + 24 = +14$$

$$A_3 = +\begin{vmatrix} 2 & 1 \\ 4 & 2 \end{vmatrix} = +(4 - 4) = 0$$

$$B_1 = +\begin{vmatrix} 5 & 7 \\ 2 & 5 \end{vmatrix} = -(25 - 14) = -11$$

$$B_2 = +\begin{vmatrix} 3 & 7 \\ 4 & 5 \end{vmatrix} = +(15 - 28) = -13$$

$$B_3 = -\begin{vmatrix} 3 & 5 \\ 4 & 2 \end{vmatrix} = -(6 - 20) = +14$$

$$C_1 = +\begin{vmatrix} 5 & 7 \\ 1 & 6 \end{vmatrix} = +(30 - 7) = +23$$

$$C_2 = -\begin{vmatrix} 3 & 7 \\ 2 & 6 \end{vmatrix} = -(18 - 14) = -4$$

$$C_3 = +\begin{vmatrix} 3 & 5 \\ 2 & 1 \end{vmatrix} = +(3 - 10) = -7$$

(*iii*) The matrix of cofactors is:

$$\begin{bmatrix} -7 & 14 & 0 \\ -11 & -13 & +14 \\ 23 & -4 & -7 \end{bmatrix}$$

$$\therefore \qquad \text{Adj. A} = \begin{bmatrix} -7 & 11 & 23 \\ -14 & -13 & +4 \\ 0 & -14 & -7 \end{bmatrix} \textbf{ Ans.}$$

Example 3.9. *Find the adjoint of the matrix*

$$A = \begin{bmatrix} 1 & 1 & 1 \\ 1 & 2 & -3 \\ 2 & -1 & 3 \end{bmatrix}$$

Solution: The cofactors of the elements of the first row are:

$$+\begin{vmatrix} 2 & -3 \\ -1 & 3 \end{vmatrix} = (2 \times 3 - 1 \times 3) = 3$$

$$-\begin{vmatrix} 1 & -3 \\ 2 & 3 \end{vmatrix} = -(3 \times 1 + 2 \times 3) = -9$$

$$+\begin{vmatrix} 1 & 2 \\ 2 & -1 \end{vmatrix} = (-1 - 2 \times 2) = -5$$

The cofactors of the elements of the second row are

$$-\begin{vmatrix} 1 & 1 \\ -1 & 3 \end{vmatrix} = -(1 \times 3 + 1 \times 1) = -4$$

$$+\begin{vmatrix} 1 & 1 \\ 2 & 3 \end{vmatrix} = +(1 \times 3 - 2 \times 1) = 1$$

$$-\begin{vmatrix} 1 & 1 \\ 2 & -1 \end{vmatrix} = -(1 \times (-1) - 2 \times 1) = 3$$

The cofactors of the elements of the third row are:

$$+\begin{vmatrix} 1 & 1 \\ 2 & -3 \end{vmatrix} = +(1 \times (-3) - 1 \times 2) = -5$$

$$-\begin{vmatrix} 1 & 1 \\ 1 & -3 \end{vmatrix} = -(1 \times (-3) - 1 \times 1) = +4$$

$$+\begin{vmatrix} 1 & 1 \\ 1 & 2 \end{vmatrix} = +(1 \times 2 - 1 \times 1) = 1$$

The matrix of cofactors is

$$\begin{bmatrix} 3 & -9 & -5 \\ -4 & 1 & 3 \\ -5 & 4 & 1 \end{bmatrix}$$

$\therefore$ Adjoint A = Transpose of the matrix of cofactors

$$= \begin{bmatrix} 3 & -4 & -5 \\ -9 & 1 & 4 \\ -5 & 3 & 1 \end{bmatrix} \quad \textbf{Ans.}$$

Example 3.10. *Find the adj. A of the matrix*

$$A = \begin{bmatrix} 1 & 0 & -1 \\ 3 & 4 & 5 \\ 0 & -6 & -7 \end{bmatrix}$$

Solution: Given matrix is

$$A = \begin{bmatrix} 1 & 0 & -1 \\ 3 & 4 & 5 \\ 0 & -6 & -7 \end{bmatrix}$$

Cofactors of the first row are:

$$+\begin{bmatrix} 4 & 5 \\ -6 & -7 \end{bmatrix} = (-28 + 30) = 2$$

$$-\begin{bmatrix} 3 & 5 \\ 0 & -7 \end{bmatrix} = -(-21 + 0) = +21$$

$$-\begin{bmatrix} 3 & 4 \\ 0 & -6 \end{bmatrix} = +(-18 + 0) = -18$$

Cofactors of the second row are:

$$-\begin{bmatrix} 0 & -1 \\ 6 & -7 \end{bmatrix} = -(0 - 6) = 6$$

$$+\begin{bmatrix} 1 & -1 \\ 0 & -7 \end{bmatrix} = +(-7 + 0) = -7$$

$$-\begin{bmatrix} 1 & 0 \\ 0 & -6 \end{bmatrix} = -(-6 + 0) = 6$$

Cofactors of the third row are:

$$+\begin{bmatrix} 0 & -1 \\ 4 & 5 \end{bmatrix} = (5 \times 0 + 4) = 4$$

$$-\begin{bmatrix} 1 & -1 \\ 3 & 5 \end{bmatrix} = -(5 + 3) = -8$$

$$+\begin{bmatrix} 1 & 0 \\ 3 & 4 \end{bmatrix} = (4 + 0) = 4$$

The matrix of cofactors is:

$$\begin{bmatrix} 2 & 21 & -18 \\ 4 & -7 & 6 \\ 4 & -8 & 4 \end{bmatrix}$$

$$\therefore \qquad \text{Adjoing A} = \begin{bmatrix} 2 & 6 & 4 \\ 21 & -7 & -8 \\ -18 & 6 & 4 \end{bmatrix} \quad \textbf{Ans.}$$

Example 3.11. *Show that the matrix*

$$\begin{bmatrix} -4 & -3 & -3 \\ 1 & 0 & 1 \\ 4 & 4 & 3 \end{bmatrix} \text{ is its own adjoint.}$$

Solution: Given matrix

$$A = \begin{bmatrix} -4 & -3 & -3 \\ 1 & 0 & 1 \\ 4 & 4 & 3 \end{bmatrix}$$

Cofactors of the first row are:

$$+\begin{bmatrix} 0 & 1 \\ 4 & 3 \end{bmatrix} = +(0 \times 3 - 4 \times 1) = -4$$

$$-\begin{bmatrix} 1 & 1 \\ 4 & 3 \end{bmatrix} = -(1 \times 3 - 4 \times 1) = +1$$

$$+\begin{bmatrix} 1 & 0 \\ 4 & 4 \end{bmatrix} = +(1 \times 4 - 4 \times 0) = +4$$

Cofactors of the second row are:

$$-\begin{bmatrix} 3 & -3 \\ 4 & 3 \end{bmatrix} = -(-3 \times 3 - (4 \times -3) = -3$$

$$+\begin{bmatrix} -4 & -3 \\ 4 & 3 \end{bmatrix} = +(-4 \times 3 + 4 \times 3) = 0$$

$$-\begin{bmatrix} -4 & -3 \\ 4 & 4 \end{bmatrix} = -(-4 \times 4 + 3 \times 4) = +4$$

Cofactors of the third row are:

$$+\begin{bmatrix} -3 & -3 \\ 0 & 1 \end{bmatrix} = +(-3 \times 1 + 3 \times 0) = -3$$

$$-\begin{bmatrix} -4 & -3 \\ 1 & 1 \end{bmatrix} = -(-4 \times 1 + 3 \times 1) = 1$$

$$+\begin{bmatrix} -4 & -3 \\ 1 & 0 \end{bmatrix} = +(-4 \times 0 + 3 \times 1) = 3$$

The matrix of cofactors is

$$\begin{bmatrix} -4 & +1 & +4 \\ -3 & 0 & 4 \\ -3 & 1 & 3 \end{bmatrix}$$

$$\therefore \qquad \text{Adjoint A} = \begin{bmatrix} -4 & -3 & -3 \\ 1 & 0 & 1 \\ 4 & 4 & 3 \end{bmatrix}$$

which is the same as the matrix A. **Proved.**

3.10 MATRIX INVERSION

Let A be a square matrix. If a matrix B exists such that product AB = product BA = unitary matrix then matrix B is called the *inverse* of A. Apparently, if B is the inverse of A then A is also the inverse of B.

The inverse of a non-singular matrix may be obtained by one of the following methods:

1. **By adjoint method.** The inverse of matrix A may be obtained by the following steps:

(*a*) Find the determinant matrix of the given matrix A.

(*b*) Calculate the value of the determinant | A |.

(*c*) Calculate the co-factors of the elements of various rows of the determinant | A |.

(*d*) Form a matrix of cofactors of determinant | A |.

(*e*) Obtain the adj. A by changing the rows into columns.

(*f*) The required inverse of the matrix A is obtained by dividing the Adj. A by the value of the determinant.

Example 3.12. *Find the inverse of*

$$A = \begin{bmatrix} 1 & 1 \\ 1 & \frac{3}{4} \end{bmatrix}$$

Solution: The determinant of the matrix A

$$= |A| = \begin{bmatrix} 1 & 1 \\ 1 & \frac{3}{4} \end{bmatrix} = 1 \times \frac{3}{4} - 1 = -\frac{1}{4}$$

Co-factor of the determinant | A | are:

$$+\frac{3}{4}, -1, -1, +1$$

∴ The matrix of co-factors of the determinant

$$= \begin{bmatrix} \frac{3}{4} & -1 \\ -1 & 1 \end{bmatrix}$$

∴ The Adjoint of A of the matrix of co-factors

$$= \begin{bmatrix} \frac{3}{4} & -1 \\ -1 & 1 \end{bmatrix}$$

∴ Inverse $A^{-1} = \frac{1}{|A|}\begin{bmatrix} \frac{3}{4} & -1 \\ -1 & 1 \end{bmatrix} = \frac{1}{-\frac{1}{4}}\begin{bmatrix} \frac{3}{4} & -1 \\ -1 & 1 \end{bmatrix}$

$$= -4\begin{bmatrix} \frac{3}{4} & -1 \\ -1 & 1 \end{bmatrix} \quad \textbf{Ans.}$$

Check: $AA^{-1} = I$

∴ $-4\begin{bmatrix} 1 & 1 \\ 1 & \frac{3}{4} \end{bmatrix} \times \begin{bmatrix} \frac{3}{4} & -1 \\ -1 & 1 \end{bmatrix}$

$$= -4\begin{bmatrix} \frac{3}{4} - 1 & -1 + 1 \\ \frac{3}{4} - \frac{3}{4} & -1 + \frac{3}{4} \end{bmatrix}$$

$$= -4\begin{bmatrix} -\frac{1}{4} & 0 \\ 0 & -\frac{1}{4} \end{bmatrix}$$

$$= \begin{bmatrix} 1 & 0 \\ 0 & 0 \end{bmatrix} = I \quad \textbf{O.K.}$$

Example 3.13. *Find the inverse of the matrix. If*

$$A = \begin{bmatrix} 2 & 2 & 4 \\ 1 & 3 & 2 \\ 3 & 1 & 3 \end{bmatrix}$$

Solution: Given matrix

$$\text{A} = \begin{bmatrix} 2 & 2 & 4 \\ 1 & 3 & 2 \\ 3 & 1 & 3 \end{bmatrix}$$

(*i*) The determinant of the matrix A = | A | = $\begin{bmatrix} 2 & 2 & 4 \\ 1 & 3 & 2 \\ 3 & 1 & 3 \end{bmatrix}$

(*ii*) The value of determinant | A |

$$= 2(3 \times 3 - 1 \times 2) - 2(1 \times 3 - 3 \times 2) + 4(1 \times 1 - 3 \times 3)$$
$$= 2 \times 7 - 2 \times (-3) + 4 \times (-8)$$
$$= 14 + 6 - 32 = -12$$

(*iii*) Cofactors of the determinant | A | are:

$$(3 \times 3 - 1 \times 2) - (1 \times 3 - 3 \times 2) + (1 \times 1 - 3 \times 3)$$
$$-(2 \times 3 - 1 \times 4) + (2 \times 3 - 3 \times 4) - (2 \times 1 - 2 \times 3)$$
$$+(2 \times 2 - 3 \times 4) - (2 \times 2 - 1 \times 4) + (2 \times 3 - 1 \times 2)$$

∴ The matrix of cofactors of the determinant, *i.e.*,

$$\begin{bmatrix} 7 & 3 & -8 \\ -2 & -6 & 4 \\ -8 & 0 & 4 \end{bmatrix}$$

(*iv*) The adj. A of the matrix of cofactors is

$$\begin{bmatrix} 7 & -2 & -8 \\ 3 & -6 & 0 \\ -8 & 4 & 4 \end{bmatrix}$$

∴ Inverse of matrix

$$\text{A} = \frac{1}{|\text{A}|}\begin{bmatrix} 7 & -2 & -8 \\ 3 & -6 & 0 \\ -8 & 4 & 4 \end{bmatrix}$$

$$= \frac{1}{-12}\begin{bmatrix} 7 & -2 & -8 \\ 3 & -6 & 0 \\ -8 & 4 & 4 \end{bmatrix} = \begin{bmatrix} \frac{-7}{12} & \frac{1}{6} & \frac{2}{3} \\ -\frac{1}{4} & \frac{1}{2} & 0 \\ \frac{2}{3} & -\frac{1}{3} & -\frac{1}{3} \end{bmatrix}$$

Check: $A \times A^{-1} = I$

$$\begin{bmatrix} 2 & 2 & 4 \\ 1 & 3 & 2 \\ 3 & 1 & 3 \end{bmatrix} \times \begin{bmatrix} \frac{-7}{12} & \frac{1}{6} & \frac{2}{3} \\ -\frac{1}{4} & \frac{1}{2} & 0 \\ \frac{2}{3} & -\frac{1}{3} & -\frac{1}{3} \end{bmatrix}$$

$$\begin{bmatrix} 2\times\left(-\frac{7}{12}\right)+2\left(-\frac{1}{4}\right)+4\times\frac{2}{3} & 2\times\frac{1}{5}+2\times\frac{1}{2}-\frac{4}{3} & 2\times\frac{2}{3}+0-\frac{4}{3} \\ 1\times\left(-\frac{7}{12}\right)+3\left(-\frac{1}{4}\right)+2\times\frac{2}{3} & 1\times\frac{1}{5}+3\times\frac{1}{2}-\frac{2}{3} & 1\times\frac{2}{3}+3\times 0-\frac{2}{3} \\ 3\times\left(-\frac{7}{12}\right)+1\left(-\frac{1}{4}\right)+3\times\frac{2}{3} & 3\times\frac{1}{6}+1\times\frac{1}{2}+3\left(-\frac{1}{3}\right) & 3\times\frac{2}{3}+1\times 0+-3\times\frac{1}{3} \end{bmatrix}$$

$$= \begin{bmatrix} 1 & 0 & 0 \\ 0 & 1 & 0 \\ 0 & 0 & 1 \end{bmatrix} = \text{A unitary matrix}$$

Example 3.14. *If* $A = \begin{bmatrix} 3 & -3 & 4 \\ 2 & -3 & 4 \\ 0 & -1 & 1 \end{bmatrix}$ *show that* $A^3 = A^{-1}$.

Solution:

$$A = \begin{bmatrix} 3 & -3 & 4 \\ 2 & -3 & 4 \\ 0 & -1 & 1 \end{bmatrix}$$

$$A^2 = \begin{bmatrix} 3 & -3 & 4 \\ 2 & -3 & 4 \\ 0 & -1 & 1 \end{bmatrix} \times \begin{bmatrix} 3 & -3 & 4 \\ 2 & -3 & 4 \\ 0 & -1 & 1 \end{bmatrix}$$

$$= \begin{bmatrix} 9-6+0, & -9+9-4, & 12-12+4 \\ 6-6+0, & -6+9-4, & 8-12+4 \\ 0-2+0, & 0+3-1, & 0-4+1 \end{bmatrix}$$

$$= \begin{bmatrix} 3 & -4 & 4 \\ 0 & -1 & 0 \\ -2 & 2 & -3 \end{bmatrix}$$

$$A^3 = A^2 \times A = \begin{bmatrix} 3 & -4 & 4 \\ 0 & -1 & 0 \\ -2 & 2 & -3 \end{bmatrix}\begin{bmatrix} 3 & -3 & 4 \\ 2 & -3 & 4 \\ 0 & -1 & 1 \end{bmatrix}$$

$$\begin{bmatrix} 9-8+0 & -9+12-4 & 12-16+4 \\ 0-2+0 & 0+3=0 & 0-4+0 \\ -6+4+0 & 6-6+4 & -8+8-3 \end{bmatrix}$$

$$A^3 = \begin{bmatrix} 1 & -1 & 0 \\ -2 & 3 & -4 \\ -2 & 3 & -3 \end{bmatrix} \qquad \text{...(A)}$$

$$A^{-1} = \frac{1}{|A|} \text{[Adjoint A]}$$

Cofactors of the first row are:

$$+\begin{bmatrix} -3 & 4 \\ -1 & 1 \end{bmatrix} = (-3+4) = 1$$

$$-\begin{bmatrix} 2 & 4 \\ 0 & 1 \end{bmatrix} = -(2+0) = -2$$

$$+\begin{bmatrix} 2 & -3 \\ 0 & -1 \end{bmatrix} = +(-2+0) = -2$$

Cofactors of the second row are:

$$-\begin{bmatrix} -3 & 4 \\ -1 & 1 \end{bmatrix} = -(-3+4) = -1$$

$$+\begin{bmatrix} 3 & 4 \\ 0 & 1 \end{bmatrix} = +(3+0) = 3$$

$$-\begin{bmatrix} 3 & -3 \\ 0 & -1 \end{bmatrix} = -(-3+0) = +3$$

Cofactors of the third row are:

$$+\begin{bmatrix} -3 & 4 \\ -3 & 4 \end{bmatrix} = +(-12+12) = 0$$

$$-\begin{bmatrix} 3 & 4 \\ 2 & 4 \end{bmatrix} = -(12-8) = -4$$

$$+\begin{bmatrix} 3 & -3 \\ 2 & -3 \end{bmatrix} = +(-9+6) = -3$$

Matrix of cofactors

$$\begin{bmatrix} 1 & -2 & -2 \\ -1 & 3 & 3 \\ 0 & -4 & -3 \end{bmatrix}$$

$$\therefore \quad \text{Adjoint A} = \begin{bmatrix} 1 & -1 & 0 \\ -2 & 3 & -4 \\ -2 & 3 & -3 \end{bmatrix}$$

$$|A| = 3(-3+4) + 3(2+0) + 4(-2+0)$$
$$= 3 \times 1 + 3 \times 2 - 8$$
$$= 3 + 6 - 8 = 1$$

$$\therefore \quad A^{-1} = \frac{1}{|A|}\begin{bmatrix} 1 & -1 & 0 \\ -2 & 3 & -4 \\ -2 & 3 & -3 \end{bmatrix}$$

$$= \begin{bmatrix} 1 & -1 & 0 \\ -2 & 3 & -4 \\ -2 & 3 & -3 \end{bmatrix} \quad \text{...(B)}$$

Matrix A = Matrix B

$$A^3 = A^{-1}.$$ **Proved.**

Example 3.15. *Obtain by the cofactor method the inverse of the matrix*

$$\begin{bmatrix} 5 & 0 & 2 \\ 0 & 2 & -3 \\ 2 & 1 & 2 \end{bmatrix}$$

Solution:

(*i*) Given matrix $A = \begin{bmatrix} 5 & 0 & 2 \\ 0 & 2 & -3 \\ 2 & 1 & 2 \end{bmatrix}$

(*ii*) The determinant of the matrix A

$$= |A| = \begin{bmatrix} 5 & 0 & 2 \\ 0 & 2 & -3 \\ 2 & 1 & 2 \end{bmatrix}$$

$$= 5(4 + 3) - 0(0 + 6) + 2(0 - 4)$$
$$= 35 + 0 - 8 = 27$$

(*iii*) Cofactors of determinant

$$+ (4 + 3), - (0 + 6) + (0 - 4)$$
$$- (0 - 2) + (10 - 4) - (5 + 0)$$
$$\begin{vmatrix} 1 & 1 \\ 2 & 3 \end{vmatrix} - \begin{vmatrix} 1 & 1 \\ 2 & 3 \end{vmatrix} \begin{vmatrix} 1 & 1 \\ 2 & 2 \end{vmatrix}$$

$$\therefore \begin{bmatrix} 6 & -15 & 6 \\ -5 & 8 & -3 \\ 1 & -1 & 0 \end{bmatrix} \text{ is the matrix of factors}$$

$$\text{Adjoint A} = \begin{bmatrix} 6 & -5 & 1 \\ -15 & 8 & -1 \\ 6 & -3 & 0 \end{bmatrix}$$

$$\therefore \quad A^{-1} = \frac{1}{|A|} \text{Adj.}(A) = \frac{1}{-3}\begin{bmatrix} 6 & -5 & 1 \\ -15 & 8 & -1 \\ 6 & -3 & 0 \end{bmatrix} = \begin{bmatrix} -2 & \frac{5}{3} & -\frac{1}{3} \\ 5 & -\frac{8}{3} & \frac{1}{3} \\ -2 & 1 & 0 \end{bmatrix} \quad \textbf{Ans.}$$

$$\text{Check: } A \times A^{-1} = \begin{bmatrix} 1 & 1 & 1 \\ 2 & 2 & 3 \\ 1 & 4 & 9 \end{bmatrix} \times \begin{bmatrix} -2 & \frac{5}{3} & -\frac{1}{3} \\ 5 & -\frac{8}{3} & \frac{1}{3} \\ -2 & 1 & 0 \end{bmatrix}$$

$$\begin{bmatrix} -1\times 2 + 1\times 5 - 1\times 2; & 1\times\frac{5}{3} + \frac{8}{3} + 1\times 1; & -1\times\frac{1}{3} + 1\times\frac{1}{3} + 1\times 0 \\ 2\times(-2) + 2\times 5 - 3\times 2; & 2\times\frac{5}{3} - 2\times\frac{8}{3} + 3\times 1; & -2\times\frac{1}{3} + 2\times\frac{1}{3} + 3\times 0 \\ -1\times 2 + 4\times 5 - 2\times 9; & 1\times\frac{5}{3} - 4\times\frac{8}{3} + 9\times 1; & -1\times\frac{1}{3} + 4\times\frac{1}{3} + 9\times 0 \end{bmatrix}$$

$$= \begin{bmatrix} 1 & 0 & 0 \\ 0 & 1 & 0 \\ 0 & 0 & 1 \end{bmatrix} = \text{unitary matrix O.K.}$$

Example 3.16. *If* $A = \begin{bmatrix} 4 & 2 & 3 \\ 2 & 3 & 4 \\ 3 & 4 & 5 \end{bmatrix}$

Obtain the value of A^{-1} *and hence varify that*

$$AA^{-1} = A^{-1}A = I$$

Solution:

(*i*) The value of $|A| = \begin{bmatrix} 4 & 2 & 3 \\ 2 & 3 & 4 \\ 3 & 4 & 5 \end{bmatrix}$

$$= 4(3 \times 5 - 4 \times 4) - 2(2 \times \;\; - 3 \times 4) + 3(2 \times 4 - 3 \times 3)$$
$$= 4(15 - 16) - 2(10 - 12) + 3(8 - 9)$$
$$= -4 + 4 - 3 = -3$$

(*ii*) Cofactors

$$(3 \times 5 - 4 \times 4), -(2 \times 5 - 3 \times 4), +(2 \times 4 - 3 \times 3)$$
$$-(2 \times 5 - 3 \times 4), +(4 \times 5 - 3 \times 3), -(4 \times 4 - 2 \times 3)$$
$$+(2 \times 4 - 3 \times 3), -(4 \times 4 - 2 \times 3), (3 \times 4 - 2 \times 2)$$

Matrix of cafactors

or $$A_{dj}A = \begin{bmatrix} -1 & 2 & -1 \\ +2 & 11 & -10 \\ -1 & -10 & 8 \end{bmatrix}$$

$\therefore$ $$A^{-1} = \frac{1}{|A|}\begin{bmatrix} -1 & 2 & -1 \\ 2 & 11 & -10 \\ -1 & -10 & 8 \end{bmatrix} = \frac{1}{-3}\begin{bmatrix} -1 & 2 & -1 \\ 2 & 11 & -10 \\ -1 & -10 & 8 \end{bmatrix}$$

$\therefore$ $$AA^{-1} = -\frac{1}{3}\begin{bmatrix} 4 & 2 & 3 \\ 2 & 3 & 4 \\ 3 & 4 & 5 \end{bmatrix}\begin{bmatrix} -1 & 2 & -1 \\ 2 & 11 & -10 \\ -1 & -10 & 8 \end{bmatrix}$$

$$= -\frac{1}{3}\begin{bmatrix} -4+4-3 & 8+22-30 & -4-20+24 \\ -2+6-4 & 4+33-40 & -2-30+32 \\ -3+8-5 & 6+44-50 & -3-40+40 \end{bmatrix}$$

$$= -\frac{1}{3}\begin{bmatrix} -3 & 0 & 0 \\ 0 & -3 & 0 \\ 0 & 0 & -3 \end{bmatrix} = \begin{bmatrix} 1 & 0 & 0 \\ 0 & 1 & 0 \\ 0 & 0 & 1 \end{bmatrix} = I$$

Again $$A^{-1}\cdot A = -\frac{1}{3}\begin{bmatrix}-1 & 2 & -1\\ 2 & 11 & -10\\ -1 & -10 & 8\end{bmatrix}\begin{bmatrix}4 & 2 & 3\\ 2 & 3 & 4\\ 3 & 4 & 5\end{bmatrix}$$

$$= -\frac{1}{3}\begin{bmatrix}-1\times 4+2\times 2-1\times 3, & 2\times 4+11\times 2-10\times 3, & -1\times 4-2\times 10+3\times 8\\ -1\times 2+2\times 3-1\times 4, & 2\times 2+11\times 3-10\times 4, & -1\times 2-10\times 3\times 8\times 4\\ -1\times 3+2\times 4-1\times 5, & 2\times 3+11\times 4-10\times 5, & -1\times 3-10\times 4+8\times 5\end{bmatrix}$$

$$= -\frac{1}{3}\begin{bmatrix}-4+4-3 & 8+22-30 & -4-20+24\\ -2+6-4 & 4+33-40 & -2-30+32\\ -3+8-5 & 6+44-50 & -3-40+40\end{bmatrix}$$

$$= -\frac{1}{3}\begin{bmatrix}-3 & 0 & 0\\ 0 & -3 & 0\\ 0 & 0 & -3\end{bmatrix} = \begin{bmatrix}1 & 0 & 0\\ 0 & 1 & 0\\ 0 & 0 & 1\end{bmatrix} = I$$

i.e., $AA^{-1} = A^{-1}A = I$ **Proved.**

Example 3.17. *Find the inverse A^{-1} of the following matrix, using any known method:*

$$A = \begin{vmatrix}1 & 2 & 3\\ 4 & 5 & 6\\ 7 & 8 & 9\end{vmatrix}$$

and, hence find the value of the matrix product AA^{-1}.

Solution: We know that the inverse of a matrix is

$$A^{-1} = \left(\frac{1}{|A|}\right) \times \text{Adj. of the matrix.}$$

The given matrix is:

$$A = \begin{vmatrix}1 & 2 & 3\\ 4 & 5 & 6\\ 7 & 8 & 9\end{vmatrix}$$

The co-factor A_{11}, A_{12}, and A_{13} are:

$$A_{11} = \begin{vmatrix}5 & 6\\ 8 & 9\end{vmatrix} = 45 - 48 = -3$$

$$A_{12} = \begin{vmatrix}4 & 6\\ 7 & 9\end{vmatrix} = -(36-42) = 6$$

$$A_{13} = \begin{vmatrix} 4 & 5 \\ 7 & 8 \end{vmatrix} = 32 - 35 = -3$$

The value of the determinant | A |

$$= a_{11}A_{11} + a_{12}A_{12} + a_{13}A_{13}$$
$$= 1 \times (-3) + 2 \times 6 + 3 \times (-3)$$
$$= -3 + 12 - 9 = 0$$

Since the value of the determinant | A | for the given matrix is zero, A^{-1} does not exist.

Example 3.18. *Numerically find the inverse of the matrix A below, clearly showing and explaing each step in the process*

$$A = \begin{bmatrix} 2 & 2 & 4 \\ 1 & 3 & 2 \\ 3 & 1 & 2 \end{bmatrix}$$

Solution: We know that the inverse of a matrix is

$$A^{-1} = \left(\frac{1}{|A|}\right) \times \text{Adj. of A}$$

The given matrix is:

$$A = \begin{bmatrix} 2 & 2 & 4 \\ 1 & 3 & 2 \\ 3 & 1 & 2 \end{bmatrix}$$

The co-factor A_{11}, A_{12}, and A_{13} are:

$$A_{11} = \begin{bmatrix} 3 & 2 \\ 1 & 2 \end{bmatrix} = 6 - 2 = 4$$

$$A_{12} = -\begin{bmatrix} 1 & 2 \\ 3 & 2 \end{bmatrix} = -(2 - 6) = 4$$

$$A_{13} = \begin{bmatrix} 1 & 3 \\ 3 & 1 \end{bmatrix} = 1 - 9 = -8$$

The value of the determinant | A |

$$= a_{11}A_{11} + a_{12}A_{12} + a_{13}A_{13}$$
$$= 2 \times 4 + 2 \times 4 - 4 \times 8 = -16$$

Co-factors of the first row are:

$$+\begin{bmatrix} 3 & 2 \\ 1 & 2 \end{bmatrix} = 6 - 2 = 4$$

$$-\begin{bmatrix} 1 & 2 \\ 3 & 2 \end{bmatrix} = -(2-6) = 4$$

$$+\begin{bmatrix} 1 & 3 \\ 3 & 1 \end{bmatrix} = 1-9 = 8$$

Co-factors of the second row are:

$$-\begin{bmatrix} 2 & 4 \\ 1 & 2 \end{bmatrix} = 4-4 = 0$$

$$+\begin{bmatrix} 2 & 4 \\ 3 & 2 \end{bmatrix} = 4-12 = -8$$

$$-\begin{bmatrix} 2 & 2 \\ 3 & 1 \end{bmatrix} = -(2-6) = 4$$

Co-factors of the third row are:

$$+\begin{bmatrix} 2 & 4 \\ 3 & 2 \end{bmatrix} = 4-12 = -8$$

$$-\begin{bmatrix} 2 & 4 \\ 1 & 2 \end{bmatrix} = -(4-4) = 0$$

$$+\begin{bmatrix} 2 & 2 \\ 1 & 3 \end{bmatrix} = 6-2 = 4$$

Matrix of co-factors:

$$\begin{bmatrix} 4 & 4 & -8 \\ 0 & -8 & 4 \\ -8 & 0 & 4 \end{bmatrix}$$

$$\therefore \qquad \text{Adjoint A} = \begin{bmatrix} 4 & 0 & -8 \\ 4 & -8 & 0 \\ -8 & 4 & 4 \end{bmatrix}$$

$\therefore$ Inverse of the matrix

$$= -\frac{1}{16}\begin{bmatrix} 4 & 0 & -8 \\ 4 & -8 & 0 \\ -8 & 4 & 4 \end{bmatrix}$$

$$= \begin{bmatrix} -\frac{1}{4} & 0 & \frac{1}{2} \\ -\frac{1}{4} & \frac{1}{2} & 0 \\ \frac{1}{2} & -\frac{1}{4} & -\frac{1}{4} \end{bmatrix}$$

Check: $AA^{-1} = I$

$$\begin{bmatrix} 2 & 2 & 4 \\ 1 & 3 & 2 \\ 3 & 1 & 2 \end{bmatrix} \begin{bmatrix} -\frac{1}{4} & 0 & \frac{1}{2} \\ -\frac{1}{4} & \frac{1}{2} & 0 \\ \frac{1}{2} & -\frac{1}{4} & -\frac{1}{4} \end{bmatrix}$$

$$= \begin{bmatrix} -\frac{1}{2}-\frac{1}{2}+2 & 0+1-1 & 1+0-1 \\ -\frac{1}{4}-\frac{3}{4}+1 & 0+\frac{3}{2}-\frac{1}{2} & \frac{1}{2}+0-\frac{1}{2} \\ -\frac{3}{4}-\frac{1}{4}+1 & 0+\frac{1}{2}-\frac{1}{2} & \frac{3}{2}+0-\frac{1}{2} \end{bmatrix}$$

$$= \begin{bmatrix} 1 & 0 & 0 \\ 0 & 1 & 0 \\ 0 & 0 & 1 \end{bmatrix} \quad \textbf{Ans.}$$

2. Gauss-Jordan method: In this method, we take a unit matrix of the same order as that of the given matrix A and write as AI. By simultaneous row operations on AI, we will try to convert A into a unit matrix. The adjacent matrix transformed from I matrix is the required inverse of A.

Example 3.19. *Using gauss elimination method determine the inverse of the matrix*

$$\begin{bmatrix} 1 & -1 & 1 \\ 5 & -4 & 3 \\ 2 & 1 & 1 \end{bmatrix}$$

Solution: Given matrix Unit matrix

$$\begin{bmatrix} 1 & -1 & 1 \\ 5 & -4 & 3 \\ 2 & 1 & 1 \end{bmatrix} \begin{bmatrix} 1 & 0 & 0 \\ 0 & 1 & 0 \\ 0 & 0 & 1 \end{bmatrix}$$

$R_2 \to (R_2 - 5R_1);\ R_3 \to (R_3 - 2R_1)$

$$= \begin{bmatrix} 1 & -1 & 1 \\ 0 & 1 & -2 \\ 0 & 0 & 5 \end{bmatrix} \begin{bmatrix} 1 & 0 & 0 \\ -5 & 1 & 0 \\ -2 & 0 & 1 \end{bmatrix}$$

$$R_3 \to (R_3 - 3R_2)$$

$$= \begin{bmatrix} 1 & -1 & 1 \\ 0 & 1 & -2 \\ 0 & 0 & 5 \end{bmatrix} \begin{bmatrix} 1 & 0 & 0 \\ -5 & 1 & 0 \\ 13 & -3 & 1 \end{bmatrix}$$

$$R_3 \to \frac{1}{5} R_3$$

$$= \begin{bmatrix} 1 & -1 & 1 \\ 0 & 1 & -2 \\ 0 & 0 & 1 \end{bmatrix} \begin{bmatrix} 1 & 0 & 0 \\ -5 & 1 & 0 \\ \frac{13}{5} & -\frac{3}{5} & \frac{1}{5} \end{bmatrix}$$

$$R_1 \to (R_2 - R_1)$$

$$\begin{bmatrix} 1 & 0 & -1 \\ 0 & 1 & -2 \\ 0 & 0 & 1 \end{bmatrix} \begin{bmatrix} -4 & 1 & 0 \\ -5 & 1 & 0 \\ \frac{13}{5} & -\frac{3}{5} & \frac{1}{5} \end{bmatrix}$$

$$R_1 \to (R_3 + R_1)$$

$$\begin{bmatrix} 1 & 0 & 0 \\ 0 & 1 & -2 \\ 0 & 0 & 1 \end{bmatrix} \begin{bmatrix} -\frac{7}{5} & \frac{2}{5} & \frac{1}{5} \\ -5 & 1 & 0 \\ \frac{13}{5} & -\frac{3}{5} & \frac{1}{5} \end{bmatrix}$$

$$R_2 \to (R_2 + 2R_3)$$

$$\begin{bmatrix} 1 & 0 & 0 \\ 0 & 1 & 0 \\ 0 & 0 & 1 \end{bmatrix} \begin{bmatrix} -\frac{7}{5} & \frac{2}{5} & \frac{1}{5} \\ \frac{1}{5} & -\frac{1}{5} & \frac{2}{5} \\ \frac{13}{5} & \frac{3}{5} & \frac{1}{5} \end{bmatrix}$$

$$\therefore \text{ Inverse of matrix } = \frac{1}{5}\begin{bmatrix} -7 & 2 & 1 \\ 1 & -1 & 2 \\ 13 & -3 & 1 \end{bmatrix} \textbf{ Ans.}$$

Check: $AA^{-1} = I$

$$\begin{bmatrix} 1 & -1 & 1 \\ 5 & -4 & 3 \\ 2 & 1 & 1 \end{bmatrix}\begin{bmatrix} -7 & 2 & 1 \\ 1 & -1 & 2 \\ 13 & -3 & 1 \end{bmatrix}$$

$$= \frac{1}{5}\begin{bmatrix} -7-1+3 & 2+1-3 & 1-2+1 \\ -35-4+39 & 10+4-9 & 5-8+3 \\ -14+1+13 & 4-1-3 & 2+2+1 \end{bmatrix}$$

$$= \frac{1}{5}\begin{bmatrix} 5 & 0 & 0 \\ 0 & 5 & 0 \\ 0 & 0 & 5 \end{bmatrix} = \begin{bmatrix} 1 & 0 & 0 \\ 0 & 1 & 0 \\ 0 & 0 & 1 \end{bmatrix} \quad \textbf{O.K.}$$

Example 3.20. *Compute the invers of* $A = \begin{bmatrix} 4 & 3 & 3 \\ -1 & 0 & -1 \\ -4 & -4 & -3 \end{bmatrix}$

Solution: We shall perform only E-row operations on the matrix A so that it reduces to a unit matrix. The same operations on a unit matrix will give the inverse of matrix A.

Matrix A Unit matrix

$$\begin{bmatrix} 4 & 3 & 3 \\ -1 & 0 & -1 \\ -4 & -4 & -3 \end{bmatrix} : \begin{bmatrix} 1 & 0 & 0 \\ 0 & 1 & 0 \\ 0 & 0 & 1 \end{bmatrix}$$

First operation. Replace R_3 by $(R_1 + R_3)$, to get the following matrix

$$\begin{bmatrix} 4 & 3 & 3 \\ -1 & 0 & -1 \\ 0 & -1 & 0 \end{bmatrix} : \begin{bmatrix} 1 & 0 & 0 \\ 0 & 1 & 0 \\ 1 & 0 & 1 \end{bmatrix}$$

Second operation. Replace R_1 by $(R_1 + 3R_2)$

$$\begin{bmatrix} 1 & 3 & 0 \\ -1 & 0 & -1 \\ 0 & -1 & 0 \end{bmatrix} : \begin{bmatrix} 1 & 3 & 0 \\ 0 & 1 & 0 \\ 1 & 0 & 1 \end{bmatrix}$$

Third operation. Replace R_1 by $(R_1 + 3R_3)$

$$\begin{bmatrix} 1 & 0 & 0 \\ -1 & 0 & -1 \\ 0 & -1 & 0 \end{bmatrix} : \begin{bmatrix} 4 & 3 & 3 \\ 0 & 1 & 0 \\ 1 & 0 & 1 \end{bmatrix}$$

Fourth operation. Interchange Row 2 and Row 3

$$\begin{bmatrix} 1 & 0 & 0 \\ 0 & -1 & 0 \\ -1 & 0 & -1 \end{bmatrix} : \begin{bmatrix} 4 & 3 & 3 \\ 1 & 0 & 1 \\ 0 & 1 & 0 \end{bmatrix}$$

Fifth operation. Replace R_3 by $(R_3 + R_1)$

$$\begin{bmatrix} 1 & 0 & 0 \\ 0 & -1 & 0 \\ 0 & 0 & -1 \end{bmatrix} : \begin{bmatrix} 4 & 3 & 3 \\ 1 & 0 & 1 \\ 4 & 4 & 3 \end{bmatrix}$$

Sixth operation. Multiply row 2 and row 3 by (– 1)

$$\begin{bmatrix} 1 & 0 & 0 \\ 0 & 1 & 0 \\ 0 & 0 & 1 \end{bmatrix} : \begin{bmatrix} 4 & 3 & 3 \\ -1 & 0 & -1 \\ -4 & -4 & -3 \end{bmatrix}$$

Since matrix A is reduced to unit matrix, the unit matrix is reduced to the desired inverse of the matrix A.

i.e., $A^{-1}\begin{bmatrix} 4 & 3 & 3 \\ -1 & 0 & -1 \\ -4 & -4 & -3 \end{bmatrix}$ **Ans.**

Check: $\begin{bmatrix} 4 & 3 & 3 \\ -1 & 0 & -1 \\ -4 & -4 & -3 \end{bmatrix}\begin{bmatrix} 4 & 3 & 3 \\ -1 & 0 & -1 \\ -4 & -4 & -3 \end{bmatrix} = \begin{bmatrix} 1 & 0 & 0 \\ 0 & 1 & 0 \\ 0 & 0 & 1 \end{bmatrix}$ **O.K.**

Example 3.21. *By Gauss-Jordan elimination method, obtain the inverse of the matrix*

$$\begin{bmatrix} 2 & 2 & 4 \\ 1 & 3 & 2 \\ 3 & 1 & 3 \end{bmatrix}$$

Solution:

$$\begin{bmatrix} 2 & 2 & 4 & : & 1 & 0 & 0 \\ 1 & 3 & 2 & : & 0 & 1 & 0 \\ 3 & 1 & 3 & : & 0 & 0 & 1 \end{bmatrix}$$

Normalising the first and third rows, we get

$$\begin{bmatrix} 1 & 1 & 2 & : & 1/2 & 0 & 0 \\ 1 & 3 & 2 & : & 0 & 1 & 0 \\ 1 & 1/3 & 1 & : & 0 & 0 & 1/3 \end{bmatrix}$$

Replacing $R_2 \to (R_2 - R_1)$ and $R_3 \to (R_3 - R_1)$, we get

$$\begin{bmatrix} 1 & 1 & 2 & : & 1/2 & 0 & 0 \\ 1 & 2 & 0 & : & -1/2 & 1 & 0 \\ 1 & -2/3 & -1 & : & -1/2 & 0 & 1/3 \end{bmatrix}$$

Dividing the second row by 2 and multiplying the third row by $\frac{-3}{2}$, we get

$$\begin{bmatrix} 1 & 1 & 2 & : & 1/2 & 0 & 0 \\ 0 & 1 & 0 & : & -1/4 & 1/2 & 0 \\ 0 & 1 & 3/2 & : & 3/4 & 0 & -1/2 \end{bmatrix}$$

Replacing $R_3 \to (R_3 - R_2)$, we get

$$\begin{bmatrix} 1 & 1 & 2 & : & 1/2 & 0 & 0 \\ 0 & 1 & 0 & : & -1/4 & 1/2 & 0 \\ 0 & 1 & 3/2 & : & 1 & -1/2 & -1/2 \end{bmatrix}$$

Dividing the third row by 3/2, we get

$$\begin{bmatrix} 1 & 1 & 2 & 1/2 & 0 & 0 \\ 0 & 1 & 0 & -1/4 & 1/2 & 0 \\ 0 & 0 & 0 & 2/3 & -1/3 & -1/3 \end{bmatrix}$$

Replacing $R_1 \to (R_1 - R_2)$, we get

$$\begin{bmatrix} 1 & 0 & 2 & 3/2 & -1/2 & 0 \\ 0 & 1 & 0 & -1/4 & 1/2 & 0 \\ 0 & 0 & 1 & 2/3 & -1/3 & -1/3 \end{bmatrix}$$

Replacing $R_1 \to (R_1 - 2R_3)$, we get

$$\begin{bmatrix} 1 & 0 & 0 & : & -7/12 & 1/6 & 2/3 \\ 0 & 1 & 0 & : & -1/4 & 1/2 & 0 \\ 0 & 0 & 1 & : & 2/3 & -1/3 & -1/3 \end{bmatrix}$$

$$\therefore \quad \text{Inverse of A} = \begin{bmatrix} -\frac{7}{12} & \frac{1}{6} & \frac{2}{3} \\ -\frac{1}{4} & \frac{1}{2} & 0 \\ \frac{2}{3} & -\frac{1}{3} & -\frac{1}{3} \end{bmatrix} \quad \textbf{Ans.}$$

Note. We have already solved this problem by the adjoint method as Example 3.13. The result may be compared.

Example 3.22. *Compute the inverse of the matrix*

$$\begin{bmatrix} 3 & -3 & 4 \\ 2 & -3 & 4 \\ 0 & -1 & 1 \end{bmatrix}$$

by applying elementary row transformation.

Solution: Perform only E-row operations on the given matrix so that it reduces to a unit matrix and perform the same operations on a unit matrix to get the required inverse of the matrix.

$$\begin{bmatrix} 3 & -3 & 4 \\ 2 & -3 & 4 \\ 0 & -1 & 1 \end{bmatrix} : \begin{bmatrix} 1 & 0 & 0 \\ 0 & 1 & 0 \\ 0 & 0 & 1 \end{bmatrix}$$

First opration: Replace R_1 by $\frac{1}{3}$ R_1,

$$\begin{bmatrix} 1 & -1 & 4/3 \\ 2 & -3 & 4 \\ 0 & -1 & 1 \end{bmatrix} : \begin{bmatrix} 1/3 & 0 & 0 \\ 0 & 1 & 0 \\ 0 & 0 & 1 \end{bmatrix}$$

Second operation: Replace R_2 by $(R_2 - 2R_1)$

$$\begin{bmatrix} 1 & -1 & 4/3 \\ 0 & -1 & 4/3 \\ 0 & -1 & 1 \end{bmatrix} : \begin{bmatrix} 1/3 & 0 & 0 \\ -2/3 & 1 & 0 \\ 0 & 0 & 1 \end{bmatrix}$$

Third operation: Replace R_2 by $-R_2$ to get

$$\begin{bmatrix} 1 & -1 & 4/3 \\ 0 & 1 & -4/3 \\ 0 & -1 & 1 \end{bmatrix} : \begin{bmatrix} 1/3 & 0 & 0 \\ 2/3 & -1 & 0 \\ 0 & 0 & 1 \end{bmatrix}$$

Fourth operation: Replace R_1 by $(R_1 + R_2)$

$$\begin{bmatrix} 1 & 0 & 0 \\ 0 & 1 & -4/3 \\ 0 & -1 & 1 \end{bmatrix} : \begin{bmatrix} 1 & -1 & 0 \\ 2/3 & -1 & 0 \\ 0 & 0 & 1 \end{bmatrix}$$

Fifth operation: Replace R_3 by $(R_2 + R_3)$

$$\begin{bmatrix} 1 & 0 & 0 \\ 0 & 1 & -4/3 \\ 0 & 0 & -1/3 \end{bmatrix} : \begin{bmatrix} 1 & -1 & 0 \\ 2/3 & -1 & 0 \\ 2/3 & -1 & 1 \end{bmatrix}$$

Sixth operation: Replace R_3 by $-3R_3$

$$\begin{bmatrix} 1 & 0 & 0 \\ 0 & 1 & -4/3 \\ 0 & 0 & 1 \end{bmatrix} : \begin{bmatrix} 1 & -1 & 0 \\ 2/3 & -1 & 0 \\ +2 & +3 & -3 \end{bmatrix}$$

Seventh operation: Replace R_2 by $(R_2 + \frac{4}{3} R_3)$

$$\begin{bmatrix} 1 & 0 & 0 \\ 0 & 1 & 0 \\ 0 & 0 & 1 \end{bmatrix} : \begin{bmatrix} 1 & -1 & 0 \\ -2 & 3 & -4 \\ -2 & 3 & -3 \end{bmatrix}$$

$\therefore \begin{bmatrix} 1 & -1 & 0 \\ -2 & 3 & -4 \\ -2 & 3 & -3 \end{bmatrix}$ is the required inverse of given matrix. **Ans.**

Check: $AA^{-1} = \begin{bmatrix} 3 & -3 & 4 \\ 2 & -3 & 4 \\ 0 & -1 & 1 \end{bmatrix} \begin{bmatrix} 1 & -1 & 0 \\ -2 & 3 & -4 \\ -2 & 3 & -3 \end{bmatrix}$

$$= \begin{bmatrix} 3+6-8 & -3-9+12 & 0+12-12 \\ 2+6-8 & -2-9+12 & 0+12-12 \\ 0+2-2 & 0-3+3 & 0+4-3 \end{bmatrix}$$

$$= \begin{bmatrix} 1 & 0 & 0 \\ 0 & 1 & 0 \\ 0 & 0 & 1 \end{bmatrix} = I, \text{ a unitary matrix}$$

Example 3.23. *Numerically find the inverse of the matrix A below, clearly showing and explaining each step in the process.*

$$A = \begin{bmatrix} 2 & 2 & 4 \\ 1 & 3 & 2 \\ 3 & 1 & 2 \end{bmatrix}$$

Name the inversion method you have used.

Solution: Let us find out the value of | A |

$$|A| = \begin{bmatrix} 2 & 2 & 4 \\ 1 & 3 & 2 \\ 3 & 1 & 2 \end{bmatrix}$$

$$= 2(6-2) - 2(2-6) + 4(1-9)$$

$$= 8 + 8 - 32 = -16$$

As the determinant | A | is not zero, the matrix A is singular. We shall calculate the inverse of the matrix A by doing elementary row transformations on matrix A till it becomes a unit matrix. By simultaneous operations, on a unit matrix we get the inverse of the matrix A.

$$\begin{bmatrix} 2 & 2 & 4 \\ 1 & 3 & 2 \\ 3 & 1 & 2 \end{bmatrix} \begin{bmatrix} 1 & 0 & 0 \\ 0 & 1 & 0 \\ 0 & 0 & 1 \end{bmatrix}$$

Replacing R_2 to $(R_3 - R_2 - R_1)$, we get

$$\begin{bmatrix} 2 & 2 & 4 \\ 1 & 3 & 2 \\ 0 & -4 & -4 \end{bmatrix} \begin{bmatrix} 1 & 0 & 0 \\ 0 & 1 & 0 \\ -1 & -1 & 1 \end{bmatrix}$$

Replacing R_2 by $(R_2 - R_{1/2})$, we get

$$\begin{bmatrix} 2 & 2 & 4 \\ 0 & 2 & 0 \\ 0 & -4 & -4 \end{bmatrix} \begin{bmatrix} 1 & 0 & 0 \\ -\frac{1}{2} & 1 & 0 \\ -1 & -1 & 1 \end{bmatrix}$$

Replacing R_1 to $(R_1 - R_2)$, we get

$$\begin{bmatrix} 2 & 0 & 4 \\ 0 & 2 & 0 \\ 0 & -4 & -4 \end{bmatrix} \begin{bmatrix} 3/2 & -1 & 0 \\ -\frac{1}{2} & 1 & 0 \\ -1 & -1 & 1 \end{bmatrix}$$

Replacing R_3 to $(R_3 + 2R_2)$, we get

$$\begin{bmatrix} 2 & 0 & 4 \\ 0 & 2 & 0 \\ 0 & 0 & -4 \end{bmatrix} \begin{bmatrix} 3/2 & -1 & 0 \\ -\frac{1}{2} & 1 & 0 \\ -2 & 1 & 1 \end{bmatrix}$$

Replacing R_1 to $(R_1 + R_3)$, we get

$$\begin{bmatrix} 2 & 0 & 0 \\ 0 & 2 & 0 \\ 0 & 0 & -4 \end{bmatrix} \begin{bmatrix} -1/2 & 0 & 1 \\ -1/2 & 1 & 0 \\ -2 & 1 & 1 \end{bmatrix}$$

Replacing R_3 to $(-R_{3/2})$

$$\begin{bmatrix} 2 & 0 & 0 \\ 0 & 2 & 0 \\ 0 & 0 & -2 \end{bmatrix} \begin{bmatrix} -1/2 & 0 & 1 \\ -1/2 & 1 & 0 \\ -1 & -1 & -1/2 \end{bmatrix}$$

Dividing the three rows by 2, we get

$$\begin{bmatrix} 1 & 0 & 0 \\ 0 & 1 & 0 \\ 0 & 0 & 1 \end{bmatrix} \begin{bmatrix} -1/4 & 0 & 1/2 \\ -1/4 & 1/2 & 0 \\ 1/2 & -1/4 & -1/4 \end{bmatrix}$$

$\therefore$ Inverse of the matrix A

$$= \begin{bmatrix} -\frac{1}{4} & 0 & \frac{1}{2} \\ -\frac{1}{4} & \frac{1}{2} & 0 \\ \frac{1}{2} & -\frac{1}{4} & -\frac{1}{4} \end{bmatrix} \quad \textbf{Ans.}$$

Ckeck:

$$\begin{bmatrix} 2 & 2 & 4 \\ 1 & 3 & 2 \\ 3 & 1 & 3 \end{bmatrix} \begin{bmatrix} -\frac{1}{4} & 0 & \frac{1}{2} \\ -\frac{1}{4} & \frac{1}{2} & 0 \\ \frac{1}{2} & -\frac{1}{4} & -\frac{1}{4} \end{bmatrix}$$

$$= \begin{bmatrix} -\frac{1}{2}-\frac{1}{2}+2 & 0+1-1 & 1+0-1 \\ -\frac{1}{4}-\frac{3}{4}+1 & 0+\frac{3}{2}-\frac{1}{2} & \frac{1}{2}+0-\frac{1}{2} \\ -\frac{3}{4}-\frac{1}{4}+1 & 0+\frac{1}{2}-\frac{1}{2} & \frac{3}{2}+0-\frac{1}{2} \end{bmatrix}$$

$$= \begin{bmatrix} 1 & 0 & 0 \\ 0 & 1 & 0 \\ 0 & 0 & 1 \end{bmatrix} \quad \textbf{O.K.}$$

Example 3.24. *Using Gauss-elimination method, fint the inverse of the matrix*

$$\begin{bmatrix} 2 & 2 & 4 \\ 1 & 3 & 2 \\ 3 & 1 & 3 \end{bmatrix}$$

Solution:
$$\begin{bmatrix} 2 & 2 & 4 \\ 1 & 3 & 2 \\ 3 & 1 & 3 \end{bmatrix} \begin{bmatrix} 1 & 0 & 0 \\ 0 & 1 & 0 \\ 0 & 0 & 1 \end{bmatrix}$$

Replacing $R_2 \rightarrow (R_1 - 2R_2)$ and $R_3 \rightarrow (3R_2 - R_3)$, we get

$$\begin{bmatrix} 2 & 2 & 4 \\ 0 & -4 & 0 \\ 0 & 8 & 3 \end{bmatrix} \begin{bmatrix} 1 & 0 & 0 \\ 1 & -2 & 0 \\ 0 & 3 & -1 \end{bmatrix}$$

Replacing $R_3 \rightarrow (R_3 + 2R_2)$, we get

$$\begin{bmatrix} 2 & 2 & 4 & : & 1 & 0 & 0 \\ 0 & -4 & 0 & : & 1 & -2 & 0 \\ 0 & 0 & 3 & : & 2 & -1 & -1 \end{bmatrix}$$

Dividing the first row by 2, the second row by – 4, and the third row by 3, we get

$$\begin{bmatrix} 1 & 1 & 2 & : & 1/2 & 0 & 0 \\ 0 & 1 & 0 & : & -1/4 & -1/2 & 0 \\ 0 & 0 & 1 & : & 2/3 & -1/3 & -1/3 \end{bmatrix}$$

Substracting the second row from the first row, we get

$$\begin{bmatrix} 1 & 0 & 2 & : & 3/4 & -1/2 & 0 \\ 0 & 1 & 0 & : & -1/4 & 1/2 & 0 \\ 0 & 0 & 1 & : & 2/3 & -1/3 & -1/3 \end{bmatrix}$$

Replacing the first row by $(R_1 - 2R_3)$.

$$\begin{bmatrix} 1 & 0 & 0 & : & -7/12 & 1/6 & 2/3 \\ 0 & 1 & 0 & : & -1/4 & 1/2 & 0 \\ 0 & 0 & 1 & : & 2/3 & -1/3 & -1/3 \end{bmatrix}$$

The required inverse of the matrix A is:

$$\begin{bmatrix} -7/12 & 1/6 & 2/3 \\ -1/4 & 1/2 & 0 \\ 2/3 & -1/3 & -1/3 \end{bmatrix} \quad \textbf{Ans.}$$

4. Short Cut method of inversion of matrix. In this method, the given matrix is transformed into its inverse by means of a unit matrix of the same order as explaxined under:

Augment the given matrix by the first column of the unit matrix and reduce to zero all the elements in the first column except one which is to be reduced to unity by dividing its row by such a number as will make it unity. Similarly, augment the transformed matrix by the second column of the unit matrix and reduce to zero all the elements in the first column except one which is further reduced to unity by dividing the elements of the row by a suitable number. The method may be studied by following the solved example.

Example 3.25. *Find the inverse of the matrix*

$$A = \begin{bmatrix} 2 & -2 & 4 \\ 2 & 3 & 2 \\ 1 & -1 & 1 \end{bmatrix}$$

Solution: Let us augment the matrix by the first column of unit matrix of the third order, *i.e.,*

$$\begin{bmatrix} 2 & -2 & 4 & : & 1 \\ 2 & 3 & 2 & : & 0 \\ 1 & -1 & 1 & : & 0 \end{bmatrix}$$

First operation. divide the first row throughout by 2

$$\begin{bmatrix} 1 & -1 & 2 & . & 1/2 \\ 2 & 3 & 2 & : & 0 \\ 1 & -1 & 1 & : & 0 \end{bmatrix}$$

Second operation. Replace R_2 by $(R_2 - 2R_1)$

$$\begin{bmatrix} 1 & -1 & 2 & : & 1/2 \\ 2 & 5 & -2 & : & -1 \\ 1 & -1 & 1 & : & 0 \end{bmatrix}$$

Third operation. Replace R_3 by $(R_3 - R_1)$

$$\begin{bmatrix} 1 & -1 & 2 & : & 1/2 \\ 0 & 5 & -2 & : & -1 \\ 0 & 0 & -1 & : & 1/2 \end{bmatrix}$$

As tht first column of the matrix has been replaced by the first column of unit matrix, discard the first column of this matrix and augment the last three columns by the second column of the unit matrix.

$$\begin{bmatrix} -1 & 2 & 1/2 & : & 0 \\ 5 & -2 & -1 & : & 1 \\ 0 & -1 & -1/2 & : & 0 \end{bmatrix}$$

First operation. Divide R_2 by 5

$$\begin{bmatrix} -1 & 2 & 1/2 & : & 0 \\ 1 & -2/5 & -1/5 & : & 1/5 \\ 0 & -1 & -1/2 & : & 0 \end{bmatrix}$$

Second operation. Replace R_1 by $(R_1 + R_2)$

$$\begin{bmatrix} 0 & 8/5 & 3/10 & : & 1/5 \\ 1 & -2/5 & -1/5 & : & 1/5 \\ 0 & -1 & -1/2 & : & 0 \end{bmatrix}$$

Now, discard the first column of the matrix and augment the resulting matrix by the third column of unit matrix.

$$\begin{bmatrix} 8/5 & 3/10 & 1/5 & : & 0 \\ -2/5 & -1/5 & 1/5 & : & 0 \\ -1 & -1/2 & 0 & : & 1 \end{bmatrix}$$

First operation. Replace R_1 by $R_1 + 8/5\ R_3$

$$\begin{bmatrix} 0 & 1/2 & 1/5 & : & 8/5 \\ -2/5 & -1/5 & 1/5 & : & 0 \\ -1 & -1/2 & 0 & : & 1 \end{bmatrix}$$

Second operation. Replace R_2 by $(R_2 - 2/3\ R_3)$

$$\begin{bmatrix} 0 & 1/2 & 1/5 & : & 0 \\ 0 & 0 & 1/5 & : & -2/5 \\ -1 & -1/2 & 0 & : & 1 \end{bmatrix}$$

Third operation. Replace R_1 by R_3 by $-R_3$

$$\begin{bmatrix} 0 & -1/2 & 1/5 & : & 8/5 \\ 0 & 0 & 1/5 & : & -2/5 \\ 1 & -1/2 & 0 & : & -1 \end{bmatrix}$$

$\therefore$ The inverse of the matrix $= \begin{bmatrix} -1/2 & 1/5 & 8/5 \\ 0 & 0 & -2/5 \\ 1/2 & 0 & -1 \end{bmatrix}$ **Ans.**

Check: $$AA^{-1} = \begin{bmatrix} 2 & -2 & 4 \\ 2 & 3 & 2 \\ 1 & -1 & 1 \end{bmatrix} \begin{bmatrix} -\frac{1}{2} & \frac{1}{5} & \frac{8}{5} \\ 0 & \frac{1}{5} & -\frac{2}{5} \\ \frac{1}{2} & 0 & -1 \end{bmatrix}$$

$$= \begin{bmatrix} 2\times\left(-\frac{1}{2}\right)+0+4\times\frac{1}{2} & 2\times\frac{1}{5}-2\times\frac{1}{5}+0 & 2\times\frac{8}{5}+2\times\frac{2}{5}-4 \\ 2\times\left(-\frac{1}{2}\right)+0+2\times\frac{1}{2} & 2\times\frac{1}{5}+3\times\frac{1}{5}+0 & 2\times\frac{8}{5}-3\times\frac{3}{5}-2 \\ 1\times\left(-\frac{1}{2}\right)+0+\frac{1}{2} & 1\times\frac{1}{5}-1\times\frac{1}{5}+0 & 1\times\frac{8}{5}+\frac{2}{5}-1 \end{bmatrix}$$

$$= \begin{bmatrix} 1 & 0 & 0 \\ 0 & 1 & 0 \\ 0 & 0 & 1 \end{bmatrix} = I.$$ **O.K.**

Example 3.26. *Obtain the inverse of the matrix*

$$A = \begin{bmatrix} 2 & 1 & 3 \\ 1 & 3 & 2 \\ 2 & 1 & 4 \end{bmatrix}$$

Solution: Augment the matrix by a unitary matrix

$$\begin{bmatrix} 2 & 1 & 3 \\ 1 & 3 & 2 \\ 2 & 1 & 4 \end{bmatrix} \begin{bmatrix} 1 & 0 & 0 \\ 0 & 1 & 0 \\ 0 & 0 & 1 \end{bmatrix}$$

Divide the first row R_1 by 2 and row R_3 by 2

$$\begin{bmatrix} 1 & \frac{1}{2} & \frac{3}{2} & : & \frac{1}{2} & 0 & 0 \\ 1 & 3 & 2 & : & 0 & 1 & 0 \\ 1 & \frac{1}{2} & 2 & : & 0 & 0 & \frac{1}{2} \end{bmatrix}$$

$R_2 \to (R_1 - R_2)$; $R_3 = (R_2 - R_3)$

$$\begin{bmatrix} 1 & \frac{1}{2} & \frac{3}{2} & : & \frac{1}{2} & 0 & 0 \\ 0 & 2\frac{1}{2} & \frac{1}{2} & : & -\frac{1}{2} & 1 & 0 \\ 0 & 2\frac{1}{2} & 0 & : & 0 & 1 & -\frac{1}{2} \end{bmatrix}$$

Multiply R_2 by 2/5; subtract R_3 from R_2 to get R_3

$$\begin{bmatrix} 1 & \frac{1}{2} & \frac{3}{2} & : & \frac{1}{2} & 0 & 0 \\ 0 & 1 & \frac{1}{5} & : & -\frac{1}{5} & \frac{2}{5} & 0 \\ 0 & 0 & \frac{1}{2} & : & -\frac{1}{2} & 0 & \frac{1}{2} \end{bmatrix}$$

Multiply R_3 by 2 to get R_3

$$\begin{bmatrix} 1 & \frac{1}{2} & \frac{3}{2} & \frac{1}{2} & 0 & 0 \\ 0 & 1 & \frac{1}{5} & -\frac{1}{5} & \frac{2}{5} & 0 \\ 0 & 0 & 1 & -1 & 0 & 1 \end{bmatrix}$$

Subtract $\frac{1}{2}$ R_2 from R_1 to get R_1

$$\begin{bmatrix} 1 & 0 & \frac{7}{5} & \frac{3}{5} & -\frac{1}{5} & 0 \\ 0 & 1 & \frac{1}{5} & -\frac{1}{5} & \frac{2}{5} & 0 \\ 0 & 0 & 1 & -1 & 0 & 1 \end{bmatrix}$$

Subtract $\frac{7}{5}$ R_3 from R_1 to get R_1

$$\begin{bmatrix} 1 & 0 & 0 & 2 & -\frac{1}{5} & -\frac{7}{5} \\ 0 & 1 & \frac{1}{5} & -\frac{1}{5} & \frac{2}{5} & 0 \\ 0 & 0 & 1 & -1 & 0 & 1 \end{bmatrix}$$

Subtract $\frac{R_3}{5}$ from R_2 to get R_2

$$\begin{bmatrix} 1 & 0 & 0 & 2 & -\frac{1}{5} & -\frac{7}{5} \\ 0 & 1 & 0 & 0 & \frac{2}{5} & -\frac{2}{5} \\ 0 & 0 & 1 & -1 & 0 & 1 \end{bmatrix}$$

$$\therefore \text{ The required inverse } = \begin{bmatrix} 2 & -\frac{2}{5} & -\frac{7}{5} \\ 0 & \frac{2}{5} & -\frac{1}{5} \\ -1 & 0 & 1 \end{bmatrix} \textbf{ Ans.}$$

Check: $AA^{-1} = I$

$$= \begin{bmatrix} 2 & 1 & 3 \\ 1 & 3 & 2 \\ 2 & 1 & 4 \end{bmatrix} \begin{bmatrix} 2 & -\frac{1}{5} & -\frac{7}{5} \\ 0 & \frac{2}{5} & -\frac{1}{5} \\ -1 & 0 & 1 \end{bmatrix}$$

$$= \begin{bmatrix} 4+0-3 & 2+0-2 & 4+0+-4 \\ -\frac{2}{5}+\frac{2}{5}+0 & -\frac{1}{5}+\frac{6}{5}+0 & -\frac{2}{5}+\frac{2}{5}+0 \\ -\frac{14}{5}-\frac{1}{5}+3 & -\frac{7}{5}-\frac{3}{5}+2 & -\frac{14}{5}-\frac{1}{5}+4 \end{bmatrix}$$

$$= \begin{bmatrix} 1 & 0 & 0 \\ 0 & 1 & 0 \\ 0 & 0 & 1 \end{bmatrix} = I \textbf{ O.K.}$$

3.11 SOLUTION OF SIMULTANEOUS EQUATIONS BY MATRIX INVERSION METHOD

The matrix of the coefficients x, y, z may be reduced to Echelon form by elementary row transformations. At the end of row transformation, the value of z may be calculated and by back substitution, the values of y and x respectively be calculated.

Example 3.27. *Solve the following equations using the matrix in version method:*

$$2x + 3y + z = 9$$
$$x + 2y + 3z = 6$$
$$3x + y + 2z = 8$$

Solution: The given equations in the matrix form, may be written as under:

$$\begin{bmatrix} 2 & 3 & 1 \\ 1 & 2 & 3 \\ 3 & 1 & 2 \end{bmatrix} \begin{bmatrix} x \\ y \\ z \end{bmatrix} = \begin{bmatrix} 9 \\ 6 \\ 8 \end{bmatrix}$$

i.e. $AX = B$

where $A = \begin{bmatrix} 2 & 3 & 1 \\ 1 & 2 & 3 \\ 3 & 1 & 2 \end{bmatrix}; X = \begin{bmatrix} x \\ y \\ z \end{bmatrix}; B = \begin{bmatrix} 9 \\ 6 \\ 8 \end{bmatrix}$

For finding the inverse of the matrix

$$A = \begin{bmatrix} 2 & 3 & 1 \\ 1 & 2 & 3 \\ 3 & 1 & 2 \end{bmatrix} \text{ proced as under:}$$

Determinant of matrix

$$|A| = \begin{vmatrix} 2 & 3 \\ 1 & 2 \end{vmatrix} - 3\begin{vmatrix} 1 & 3 \\ 3 & 2 \end{vmatrix} + \begin{vmatrix} 1 & 2 \\ 3 & 1 \end{vmatrix}$$

$$= 2 - 3(-7) - 5 = 18$$

Matrix of cofactors:

$$\begin{matrix} \begin{vmatrix} 2 & 3 \\ 1 & 2 \end{vmatrix} & -\begin{vmatrix} 1 & 3 \\ 3 & 2 \end{vmatrix} & \begin{vmatrix} 1 & 2 \\ 3 & 1 \end{vmatrix} \\ -\begin{vmatrix} 3 & 1 \\ 1 & 2 \end{vmatrix} & \begin{vmatrix} 2 & 1 \\ 3 & 2 \end{vmatrix} & -\begin{vmatrix} 2 & 3 \\ 3 & 1 \end{vmatrix} \\ \begin{vmatrix} 3 & 1 \\ 2 & 3 \end{vmatrix} & -\begin{vmatrix} 2 & 1 \\ 1 & 3 \end{vmatrix} & \begin{vmatrix} 2 & 3 \\ 1 & 2 \end{vmatrix} \end{matrix}$$

i.e.,

$$\begin{bmatrix} 1 & 7 & -5 \\ -5 & 1 & -7 \\ 7 & -6 & 1 \end{bmatrix}$$

Transposing the matrix of cofactors

$$\text{Adj. }(A) = \begin{bmatrix} 1 & -5 & 7 \\ 7 & 1 & -5 \\ -5 & 7 & 1 \end{bmatrix}$$

$$\therefore \quad A^{-1} = \frac{1}{|A|}\begin{bmatrix} 1 & -5 & 7 \\ 7 & 1 & -5 \\ -5 & 7 & 1 \end{bmatrix}$$

$$X = A^{-1}B$$

$$= \frac{1}{18}\begin{bmatrix} 1 & -5 & 7 \\ 7 & 1 & -5 \\ -5 & 7 & 1 \end{bmatrix}\begin{bmatrix} 9 \\ 6 \\ 8 \end{bmatrix}$$

$$X = \frac{1}{18}\begin{bmatrix} 9 & -30 & 56 \\ 63 & 6 & -40 \\ -45 & 42 & 8 \end{bmatrix} = \frac{1}{18}\begin{bmatrix} 35 \\ 29 \\ 5 \end{bmatrix}$$

$$\begin{bmatrix} x \\ y \\ z \end{bmatrix} = \frac{1}{18}\begin{bmatrix} 35 \\ 29 \\ 5 \end{bmatrix}$$

$$\therefore \quad x = \frac{35}{18}; y = \frac{29}{18}; z = \frac{5}{18} \quad \textbf{Ans.}$$

Example 3.28. *Solve the following equations by inversion method*

$$x + y + z = 3$$
$$x + 2y + 3z = 4$$
$$x + 4y + 9z = 6$$

Solution: The given equations in the matrix form are written as under:

$$\begin{bmatrix} 1 & 1 & 1 \\ 1 & 2 & 3 \\ 1 & 4 & 9 \end{bmatrix}, \begin{bmatrix} x \\ y \\ z \end{bmatrix} = \begin{bmatrix} 3 \\ 4 \\ 6 \end{bmatrix}$$

i.e., $\quad AX = B$

where $\quad A = \begin{bmatrix} 1 & 1 & 1 \\ 1 & 2 & 3 \\ 1 & 4 & 9 \end{bmatrix}, X = \begin{bmatrix} x \\ y \\ z \end{bmatrix}, B = \begin{bmatrix} 3 \\ 4 \\ 6 \end{bmatrix}$

$$X = A^{-1} B.$$

We have to find out the inverse of matrix $A = \begin{bmatrix} 1 & 1 & 1 \\ 1 & 2 & 3 \\ 1 & 4 & 9 \end{bmatrix}$

$$|A| = 1 \times (18 - 12) - 1(9 - 3) + 1(4 - 2)$$
$$= 6 - 6 + 2 = 2$$

Cofactors:

$$= (18 - 12), - 9(9 - 3), + (4 - 2),$$
$$- (9 - 4), + (9 - 1), - (4 - 1),$$

$$+ (3, 2), - (3 - 1), + (2 - 1)$$

Now, matrix of cofactors

$$A = \begin{bmatrix} 6 & -6 & 2 \\ -5 & +8 & -3 \\ 1 & -2 & 1 \end{bmatrix}$$

$$\therefore \quad \text{Adjoing of A} = \begin{bmatrix} 6 & -5 & 1 \\ -6 & 8 & -2 \\ 2 & -3 & 1 \end{bmatrix}$$

$$A^{-1} = \frac{1}{|A|} \text{Adjoint A}$$

$$= \frac{1}{2}\begin{bmatrix} 6 & -5 & 1 \\ -6 & 8 & -2 \\ 2 & -3 & 1 \end{bmatrix}$$

$$X = A^{-1}B$$

$$= \frac{1}{2}\begin{bmatrix} 6 & -5 & 1 \\ -6 & 8 & -2 \\ 2 & -3 & 1 \end{bmatrix}\begin{bmatrix} 3 \\ 4 \\ 6 \end{bmatrix}$$

or $$\begin{bmatrix} x \\ y \\ z \end{bmatrix} = \frac{1}{2}\begin{bmatrix} 18 & -20 & 6 \\ -18 & 32 & -12 \\ 6 & -12 & 6 \end{bmatrix}$$

or $$\begin{bmatrix} x \\ y \\ z \end{bmatrix} = \frac{1}{2}\begin{bmatrix} 4 \\ 2 \\ 0 \end{bmatrix} = \begin{bmatrix} 2 \\ 1 \\ 0 \end{bmatrix}$$

$\therefore$ $x = 2, y = 1, z = 0.$ **Ans.**

Example 3.29. *Solve the following equation by inversion method*

$$4y_1 + 6y_2 - 2y_3 = 10$$
$$8y_1 + 8y_2 - 6y_3 = 6$$
$$4y_1 - 6y_2 + 2y_3 = -2$$

Solution: The matrix of coefficients is:

$$\begin{bmatrix} 4 & 6 & -2 & : & 1 & 0 & 0 \\ 8 & 8 & -6 & : & 0 & 1 & 0 \\ 4 & -6 & +2 & : & 0 & 0 & 0 \end{bmatrix}\begin{bmatrix} y_1 \\ y_2 \\ y_3 \end{bmatrix} = \begin{bmatrix} 10 \\ 6 \\ -2 \end{bmatrix}$$

Subtracting R_2 from $2R_1$ and R_3 from R_1 to get R_2 and R_3 respectively

$$= \begin{bmatrix} 4 & 6 & -2 & : & 1 & 0 & 0 \\ 0 & 4 & 2 & : & 2 & -1 & 0 \\ 0 & 12 & -4 & : & 1 & 0 & -1 \end{bmatrix}$$

Subtracting $2R_1$ from R_3, we get

$$= \begin{bmatrix} -8 & 0 & 0 & : & -1 & 0 & -1 \\ 0 & 4 & 2 & : & 2 & -1 & 0 \\ 0 & 12 & 4 & : & 1 & 0 & -1 \end{bmatrix}$$

Adding $2R_2$ and R_3 to get R_2

$$= \begin{bmatrix} -8 & 0 & 0 & : & -1 & 0 & -1 \\ 0 & 20 & 0 & : & 2 & -2 & -1 \\ 0 & 12 & -4 & : & 1 & 0 & -1 \end{bmatrix}$$

Dividing R_2 by 20 and R_3 by – 4, we get

$$= \begin{bmatrix} -8 & 0 & 0 & : & -1 & 0 & -1 \\ 0 & 1 & 0 & : & 1/4 & -1/10 & -1/20 \\ 0 & -3 & 1 & : & -1/4 & 0 & +1/4 \end{bmatrix}$$

Replacing R_3 by $R_3 + 3R_2$, we get

$$= \begin{bmatrix} -8 & 0 & 0 & : & -1 & 0 & -1 \\ 0 & 1 & 0 & : & 1/4 & -1/10 & -1/20 \\ 0 & 0 & 1 & : & 1/2 & -3/10 & +1/10 \end{bmatrix}$$

Dividing R_1 by – 8, we get

$$\begin{bmatrix} 1 & 0 & 0 & : & 1/8 & 0 & 1/8 \\ 0 & 1 & 0 & : & 1/4 & -1/10 & -1/20 \\ 0 & 0 & 1 & : & 1/2 & -3/10 & +1/10 \end{bmatrix} \begin{bmatrix} 10 \\ 6 \\ -2 \end{bmatrix}$$

$$\frac{1}{8} \times 10 + 0 \times 6 - \frac{1}{8} \times 2 = y_1$$

$$\therefore \quad y_1 = \frac{5}{4} - \frac{1}{4} = 1$$

$$\frac{1}{4} \times 10 - \frac{1}{10} \times 6 - \frac{1}{20} \times (-2) = y_2$$

$$\therefore \quad y_2 = \frac{5}{2} - \frac{3}{5} + \frac{1}{10} = 2$$

$$\frac{1}{2}\times 10-\frac{3}{10}\times 6+\frac{1}{10}(-2) = y_3$$

$$\therefore \quad y_3 = 5-\frac{9}{5}-\frac{1}{5} = 3$$

$$\therefore \quad \left.\begin{aligned} y_1 &= 1 \\ y_2 &= 2 \\ y_3 &= 3 \end{aligned}\right\} \textbf{ Ans.}$$

Example 3.30. *Solve the following equations*

$$2x_1 + 3x_2 - x_3 = 5$$
$$4x_1 + 4x_2 - 3x_3 = 3$$
$$2x_1 + 3x_2 + x_3 = -1$$

Solution: In the matrix form, the equations may be written as under

$$\begin{bmatrix} 2 & 3 & -1 \\ 4 & 4 & -3 \\ 2 & -3 & 1 \end{bmatrix}\begin{bmatrix} x_1 \\ x_2 \\ x_3 \end{bmatrix} = \begin{bmatrix} 5 \\ 3 \\ -1 \end{bmatrix}$$

For reducing the matrix into upper triangular matrix, follow the following:

Subtracting R_2 from $2R_1$, we get

$$\begin{bmatrix} 2 & 3 & -1 \\ 0 & 2 & 1 \\ 2 & -3 & 1 \end{bmatrix}\begin{bmatrix} x_1 \\ x_2 \\ x_3 \end{bmatrix} = \begin{bmatrix} 5 \\ 7 \\ -1 \end{bmatrix}$$

Subtracting R_3 from R_1, we get

$$\begin{bmatrix} 2 & 3 & -1 \\ 0 & 2 & 1 \\ 2 & 6 & -2 \end{bmatrix}\begin{bmatrix} x_1 \\ x_2 \\ x_3 \end{bmatrix} = \begin{bmatrix} 5 \\ 7 \\ 6 \end{bmatrix}$$

Subtracting R_3 from $3R_2$, we get

$$\begin{bmatrix} 2 & 3 & -1 \\ 0 & 2 & 1 \\ 0 & 0 & 5 \end{bmatrix}\begin{bmatrix} x_1 \\ x_2 \\ x_3 \end{bmatrix} = \begin{bmatrix} 5 \\ 7 \\ 15 \end{bmatrix}$$

$$\therefore \quad 2x_1 + 3x_2 - x_3 = 5 \qquad ...(i)$$
$$2x_2 + x_3 = 7 \qquad ...(ii)$$
$$5x_3 = 15 \qquad ...(iii)$$

or $$x_3 = \frac{15}{5} = 3$$

Substituting $x_3 = 3$ in eqn. (*ii*), we get

$$2x_2 + 3 = 7$$

or $$x_2 = 2$$

Substituting $x_3 = 3$, $x_2 = 2$ in eqn. (*i*), we get

$$2x_1 + 3 \times 2 - 3 = 5$$

$$2x_1 = 5 + 3 - 6$$

or $$x_1 = 1$$

Hence $$\left.\begin{array}{l} x_1 = 1 \\ x_2 = 2 \\ x_3 = 3 \end{array}\right\}$$ **Ans.**

Example 3.31. *Solve the following system of equations using Gauss — Jordan elimination method, clearly indicating each step.*

$$x_1 + 2x_2 + x_3 + 4x_4 = 13$$
$$2x_1 + 4x_3 + 3x_4 = 28$$
$$4x_1 + 2x_2 + 2x_3 + x_4 = 20$$
$$-3x_1 + x_2 + 3x_3 + 2x_4 = 6$$

Solution: Write the given equations in the form of AX = B

The given equations can be written in the matrix form as:

$$\begin{bmatrix} 1 & 2 & 1 & 4 \\ 2 & 0 & 4 & 3 \\ 4 & 2 & 2 & 1 \\ -3 & 1 & 3 & 2 \end{bmatrix} \begin{bmatrix} x_1 \\ x_2 \\ x_3 \\ x_4 \end{bmatrix} = \begin{bmatrix} 13 \\ 28 \\ 20 \\ 6 \end{bmatrix}$$

Now, we shall do the transformation in such a way that the matrix A becomes an upper triangulation matrix.

Step 1. $R_2 = (R_2 - 2R_1)$; $R_3 = (R_3 - 4R_1)$; $R_4 = (R_4 + 3R_1)$

$$\begin{bmatrix} 1 & 2 & 1 & 4 \\ 0 & -4 & 2 & -5 \\ 0 & -6 & -2 & -15 \\ 0 & 7 & 6 & 14 \end{bmatrix} \begin{bmatrix} x_1 \\ x_2 \\ x_3 \\ x_4 \end{bmatrix} = \begin{bmatrix} 13 \\ 2 \\ -32 \\ 45 \end{bmatrix}$$

Step 2. $R_3 = (R_3 - 1.5\, R_2)$; $R_4 = (R_4 + 1.75\, R_2)$

$$\begin{bmatrix} 1 & 2 & 1 & 4 \\ 0 & -4 & 2 & -5 \\ 0 & 0 & -5 & -7.5 \\ 0 & 0 & 9.5 & 5.25 \end{bmatrix} \begin{bmatrix} x_1 \\ x_2 \\ x_3 \\ x_4 \end{bmatrix} = \begin{bmatrix} 13 \\ 2 \\ -35 \\ 48.5 \end{bmatrix}$$

Step 3. $R_4 = (R_4 + 1.9 R_3)$

$$\begin{bmatrix} 1 & 2 & 1 & 4 \\ 0 & -4 & 2 & -5 \\ 0 & 0 & -5 & -7.5 \\ 0 & 0 & 0 & 9.0 \end{bmatrix} \begin{bmatrix} x_1 \\ x_2 \\ x_3 \\ x_4 \end{bmatrix} = \begin{bmatrix} 13 \\ 2 \\ -35 \\ -18 \end{bmatrix}$$

Substituting the values from right to left we get

$$\begin{aligned} -9x_4 &= -18 \\ \therefore \quad x_4 &= 2 \\ -5x_3 + 7.5x_4 &= -35 \\ 5x_3 + 7.5x_4 &= 35 \\ 5x_3 &= 35 - 7.5 \times 2 = 20 \\ x_3 &= \frac{20}{5} = 4 \\ -4x_2 + 2x_3 - 5x_4 &= 2 \\ 4x_2 - 2x_3 + 5x_4 &= -2 \\ 4x_2 &= -2 + 2 \times 4 - 5 \times 2 = -2 + 8 - 10 = -4 \\ x_2 &= -1 \\ x_1 + 2x_2 + x_3 + 4x_4 &= 13 \\ x_1 &= 13 - 2x_2 - x_3 - 4x_4 \\ &= 13 + 2 \times 1 - 4 - 4 \times 2 \\ &= 13 + 2 - 4 - 8 = 3 \\ \therefore \quad x_1 &= 1 \end{aligned}$$

Hence, $x_1 = 3; x_2 = -1; x_3 = 4; x_4 = 2$ **Ans.**

Example 3.32. *Solve the following system of equations by the elimination method due to Gauss*

$$\begin{aligned} 3w + x + 3y + z &= 6 \\ w + 2x + 2y + 3z &= -3 \\ 4w - 3x + 3y + 4z &= 5 \\ 2w - 4x - y + 2z &= 0 \end{aligned}$$

Solution: The equations may be expressed in matrix form as under:

$$\begin{bmatrix} 3 & 1 & 3 & 1 \\ 1 & 2 & 2 & 3 \\ 4 & -3 & 3 & 4 \\ 2 & -4 & -1 & 2 \end{bmatrix} \begin{bmatrix} w \\ x \\ y \\ z \end{bmatrix} = \begin{bmatrix} 6 \\ -3 \\ 5 \\ 0 \end{bmatrix}$$

By transforming:

$$\begin{bmatrix} 1 & \frac{1}{3} & 1 & \frac{1}{3} \\ 1 & 2 & 2 & 3 \\ 1 & -\frac{3}{4} & \frac{3}{4} & 1 \\ 1 & -2 & -\frac{1}{2} & 1 \end{bmatrix} \begin{bmatrix} w \\ x \\ y \\ z \end{bmatrix} = \begin{bmatrix} 2 \\ -3 \\ \frac{5}{4} \\ 0 \end{bmatrix} \begin{matrix} R_1 = \frac{1}{3}R_1 \\ R_2 = R_2 \\ R_3 = \frac{1}{4}R_3 \\ R_4 = \frac{1}{2}R_4 \end{matrix}$$

By transforming:

$$\begin{bmatrix} 1 & \frac{1}{3} & 1 & \frac{1}{3} \\ 0 & \frac{5}{3} & 1 & \frac{8}{3} \\ 0 & \frac{11}{4} & \frac{5}{4} & 2 \\ 0 & -\frac{5}{4} & -\frac{5}{4} & 0 \end{bmatrix} \begin{bmatrix} w \\ x \\ y \\ z \end{bmatrix} = \begin{bmatrix} 2 \\ -5 \\ -\frac{17}{4} \\ -\frac{5}{4} \end{bmatrix} \begin{matrix} R_2 - R_1 \\ R_2 - R_2 \\ R_4 - R_3 \end{matrix}$$

By transforming:

$$\begin{bmatrix} 1 & \frac{1}{3} & 1 & \frac{1}{3} \\ 0 & 1 & \frac{3}{5} & \frac{8}{3} \\ 0 & 1 & \frac{5}{11} & \frac{8}{11} \\ 0 & -1 & -1 & 0 \end{bmatrix} \begin{bmatrix} w \\ x \\ y \\ z \end{bmatrix} = \begin{bmatrix} 2 \\ -3 \\ -\frac{17}{11} \\ -1 \end{bmatrix} \begin{matrix} \frac{3}{5}R_2 \\ \frac{4}{11}R_3 \\ \frac{4}{5}R_4 \end{matrix}$$

By transforming:

$$\begin{bmatrix} 1 & \frac{1}{3} & 1 & \frac{1}{3} \\ 0 & 1 & \frac{3}{5} & \frac{8}{5} \\ 0 & 0 & \frac{8}{55} & \frac{48}{55} \\ 0 & 0 & -\frac{6}{11} & \frac{8}{11} \end{bmatrix} \begin{bmatrix} w \\ x \\ y \\ z \end{bmatrix} = \begin{bmatrix} 2 \\ -3 \\ -\frac{16}{11} \\ -\frac{28}{11} \end{bmatrix} \begin{matrix} R_2 - R_3 \\ R_3 + R_4 \end{matrix}$$

By transforming:

$$\begin{bmatrix} 1 & \frac{1}{3} & 1 & \frac{1}{3} \\ 0 & 1 & \frac{3}{5} & \frac{8}{5} \\ 0 & 0 & 1 & 6 \\ 0 & 0 & -1 & \frac{4}{3} \end{bmatrix} \begin{bmatrix} w \\ x \\ y \\ z \end{bmatrix} = \begin{bmatrix} 2 \\ -3 \\ -10 \\ -\frac{14}{3} \end{bmatrix} \quad \begin{matrix} R_2 = \frac{55}{8} R_3 \\ R_4 = \frac{11}{6} R_4 \end{matrix}$$

$$\begin{bmatrix} 1 & \frac{1}{3} & 1 & \frac{1}{3} \\ 0 & 1 & \frac{3}{5} & \frac{8}{5} \\ 0 & 0 & 1 & 6 \\ 0 & 0 & 0 & \frac{22}{3} \end{bmatrix} \begin{bmatrix} w \\ x \\ y \\ z \end{bmatrix} = \begin{bmatrix} 2 \\ -3 \\ -10 \\ -\frac{44}{3} \end{bmatrix}$$

$\therefore$

$$x + \frac{1}{3}x + y + \frac{1}{3}z = 2 \qquad ...(i)$$

$$x + \frac{3}{5}y + \frac{8}{5}z = -3 \qquad ...(ii)$$

$$y + 6z = -10 \qquad ...(iii)$$

$$\frac{22}{3}z = \frac{-44}{3} \qquad ...(iv)$$

From eqn. (*iv*), we get

$$z = \frac{-44}{3} \times \frac{3}{22} = -2$$

From eqn. (*iii*), we get

$$y + 6 \times (-2) = -10$$

$$y + 12 - 10 = 2$$

From eqn. (*ii*), we get

$$x + \frac{3}{5} \times 2 + \frac{8}{5} \times (-2) = -3$$

$$x = -3 - \frac{6}{5} + \frac{16}{5} = 1$$

From eqn. (*i*), we get

$$w + \frac{1}{3} \times (-1) + 2 + \frac{1}{3}(-2) = 2 \Rightarrow w = 2 - 1 = 1$$

$$\therefore \text{ Hence } \left.\begin{array}{l} w=1 \\ x=-1 \\ y=2 \\ z=-2 \end{array}\right\} \textbf{Ans.}$$

Example 3.33. *Solve the following system of linear simultaneous equations by Gaussion dimination method.*

$$\begin{aligned} x_1 + 2x_2 + x_3 + 4x_4 &= 13 \\ 2x_1 + 0x_2 + 4x_3 + 3x_4 &= 28 \\ 4x_1 + 2x_2 + 2x_3 + x_4 &= 20 \\ -3x_1 + x_2 + 3x_3 + 2x_4 &= 6 \end{aligned}$$

Solution: Applying the steps of Gaussion elimination method.

$$x_1 + 2x_2 + x_3 + 4x_4 = 13 \quad ...(1)$$
$$2x_1 + 0x_2 + 4x_3 + 3x_4 = 28 \quad ...(2)$$
$$4x_1 + 2x_2 + 2x_3 + x_4 = 20 \quad ...(3)$$
$$-3x_1 + x_2 + 3x_3 + 2x_4 = 6 \quad ...(4)$$

Step 1. For eliminating the coefficients of x_1 in eqns. (2), (3) and (4) we use.

Eqn. (2) $\to$ (eqn. (2) – 2 eqn. (1)), eqn. (3) $\to$ (eqn. (3) – 4 eqn. (1)); eqn. (4) $\to$ (eqn. (4) + 3 eqn. (1)).

$$\therefore \quad x_1 + 2x_2 + x_3 + 4x_4 = 13 \quad ...(1')$$
$$-4x_2 + 2x_3 - 5x_4 = 2 \quad ...(2')$$
$$-6x_2 - 2x_3 - 15x_4 = -32 \quad ...(3')$$
$$7x_2 + 6x_3 + 14x_4 = 45 \quad ...(4')$$

Step 2. Eq. (3″) $\to$ (eqn. (3′) – 1.5 eq. (2′)), eqn. (4″) $\to$ (eqn. (4′) + 1.75 eqn. (2′))

$$x_1 + 2x_2 + x_3 + 4x_4 = 13 \quad ...(1'')$$
$$-4x_2 + 2x_3 - 5x_4 = 2 \quad ...(2'')$$
$$-5x_2 - 7.5x_4 = -35 \quad ...(3'')$$
$$9.5x_3 - 5.25x_4 = 48.5 \quad ...(4'')$$

Step 3. Eq. (4‴) $\to$ eqn. (4″) + $1.9x_3$

$$x_1 + 2x_2 + x_3 + 4x_4 = 13 \quad ...(1''')$$
$$-4x_2 + 2x_3 - 5x_4 = 2 \quad ...(2''')$$
$$-5x_2 - 7.5x_4 = -35 \quad ...(3''')$$
$$-9x_4 = -18 \quad ...(4''')$$

Applying the backward substitution we get

$$-9x_4 = 18 \quad \text{or} \quad x_4 = 2$$
$$-5x_3 - 7.5x_4 = -35 \quad \text{or} \quad +5x_3 + 7.5 \times 2 = 35$$

$$5x_3 = 35 - 15 = 20 \quad \therefore \quad x_3 = 4$$

$$-4x_2 + 2x_3 - 5x_4 = 2 \quad \text{or} \quad 4x_2 - 2 \times 4 + 5 \times 2 = -2$$

$$x_2 = -1$$

$$x_1 + 2x_2 + x_3 + 4x_4 = 13 \quad \text{or} \quad x_1 + 2 \times (-1) + 4 + 4 \times 2 = 13$$

$$x_1 = 13 + 2 - 4 - 8 = 3$$

Hence $x_1 = 3$; $x_2 = -1$; $x_3 = 4$; $x_4 = 2$ **Ans.**

Example 3.34. *Solve the set of simultaneous equations*

$$x - y + 2z = 5$$
$$3x + y - z = 2$$
$$2x + 2y - 3z = -3$$

using matrix inversion technique.

Solution: The given system of equations are:

$$x - y + 2z = 5$$
$$3x + y - z = 2$$
$$2x + 2y - 3z = -3$$

The equation may be written in matrix form as:

$$\begin{bmatrix} 1 & -1 & 2 \\ 3 & 1 & -1 \\ 2 & 2 & -3 \end{bmatrix} \begin{bmatrix} x \\ y \\ z \end{bmatrix} = \begin{bmatrix} 5 \\ 2 \\ -3 \end{bmatrix}$$

or $$AX = D$$

Where $$A = \begin{bmatrix} 1 & -1 & 2 \\ 3 & 1 & -1 \\ 2 & 2 & -3 \end{bmatrix} X = \begin{bmatrix} x \\ y \\ z \end{bmatrix} \text{ and } D = \begin{bmatrix} 5 \\ 2 \\ -3 \end{bmatrix}$$

$$|A| = \begin{vmatrix} 1 & -1 & 2 \\ 3 & 1 & -1 \\ 2 & 2 & -3 \end{vmatrix}$$

$$= 1(-3 + 2) - (-1)(-9 + 2) + 2(6 - 2)$$

$$= -1 - 7 + 8 = 0$$

Since $|A| = 0$, A^{-1} does not exist.

Considering first two equations

$$x - y + 2z = 5 \qquad \text{...}(i)$$

$$3x + y - z = 2 \qquad \text{...}(ii)$$

Putting $z = k$ in eqn. (*i*) and (*ii*), we get

$$x - y + 2k = 5$$
$$3x + y - k = 2$$

$$\therefore \quad \begin{aligned} x - y &= 5 - 2k \\ 3x + y &= 2 + k \end{aligned} \qquad \text{...(iii)}$$

The system of equations (*iii*) can be written as

$$\begin{bmatrix} 1 & -1 \\ 3 & 1 \end{bmatrix} \begin{bmatrix} x \\ y \end{bmatrix} = \begin{bmatrix} 5-2k \\ 2+k \end{bmatrix}$$

or $$BX = D \qquad \text{...(iv)}$$

where, $$B = \begin{bmatrix} 1 & -1 \\ 3 & 1 \end{bmatrix}, X = \begin{bmatrix} x \\ y \end{bmatrix}, D = \begin{bmatrix} 5-2k \\ 2+k \end{bmatrix}$$

Now $$|B| = \begin{vmatrix} 1 & -1 \\ 3 & 1 \end{vmatrix}$$

Hence B^{-1} exists.

Let B_{ij} be the cofactor of the element in the ith row and *i*th column.

$$B_{11} = 1, B_{12} = -3, B_{21} = 1, B_{22} = 1$$

$\therefore$ Cofactor matrix $C = \begin{bmatrix} B_{11} & B_{12} \\ B_{21} & B_{22} \end{bmatrix}$

Adj. $$B = C^T = \begin{bmatrix} B_{11} & B_{12} \\ B_{21} & B_{22} \end{bmatrix} = \begin{bmatrix} 1 & 1 \\ -3 & 1 \end{bmatrix}$$

$$\therefore \quad B^{\,1} = \frac{A_{dj}B}{|B|} = \frac{1}{4}\begin{bmatrix} 1 & 1 \\ -3 & 1 \end{bmatrix}$$

From eqn. (*iv*), we get

$$X = B^{-1} D$$

$$= \frac{1}{4}\begin{bmatrix} 1 & 1 \\ -3 & 1 \end{bmatrix}\begin{bmatrix} 5-2k \\ 2+k \end{bmatrix}$$

$$= \frac{1}{4}\begin{bmatrix} 5-2k+2+k \\ -15+6k+2+k \end{bmatrix}$$

$$\begin{bmatrix} x \\ y \end{bmatrix} = \frac{1}{4}\begin{bmatrix} 7-k \\ -13+7k \end{bmatrix}$$

or $$x = \frac{1}{4}(7-k), y = \frac{1}{4}(-13+7k), z = k.$$ **Ans.**

Example 3.35. *Solve the following set of simultaneous equations using Gaussian elimination method.*

$$\begin{aligned} 3x - y + 5z &= 16 \\ x - 3y + 2z &= 1 \\ 4x + y + z &= 9 \end{aligned}$$

Solution: The given equation are:

$$\begin{aligned} 3x - y + 5z &= 16 \\ x - 3y + 2z &= 1 \\ 4x + y + z &= 9 \end{aligned}$$

The augmented coefficient matrix is:

$$\begin{bmatrix} 3 & -1 & 5 & 16 \\ 1 & -3 & 2 & 1 \\ 4 & 1 & 1 & 9 \end{bmatrix}$$

Replacing R_2 by $\left(R_2 - \frac{1}{3}R_1\right)$, we get

$$\begin{bmatrix} 3 & -1 & 5 & 16 \\ 0 & -2\frac{2}{4} & \frac{1}{3} & -4\frac{1}{3} \\ 4 & 1 & 1 & 9 \end{bmatrix}$$

Replacing R_3 by $\left(\frac{3}{4}R_3 - R_1\right)$, we get

$$\begin{bmatrix} 3 & -1 & 5 & 16 \\ 0 & -2\frac{2}{3} & \frac{1}{3} & -4\frac{1}{3} \\ 0 & 1\frac{3}{4} & -4\frac{1}{4} & -9\frac{1}{4} \end{bmatrix}$$

Replacing R_3 by $-\left(2\frac{2}{3}R_3 + 1\frac{3}{4}R_2\right)$, we get

$$\begin{bmatrix} 3 & -1 & 5 & 16 \\ 0 & -2\frac{2}{3} & \frac{1}{3} & -4\frac{1}{3} \\ 0 & 0 & 10\frac{3}{4} & 32\frac{1}{4} \end{bmatrix}$$

$\therefore \quad 3x - y + 5z = 16$

$$-2\frac{2}{3}y + \frac{1}{3}y = -4\frac{1}{3}$$

$$10\frac{3}{4}z = 32.25$$

or $$z = \frac{32.25}{10.75} = 3$$

$$y = -\frac{3}{8}\left(\frac{13}{3} - \frac{1}{3} \times 3\right)$$

$$= -\frac{3}{8}\left(-\frac{13}{3} - 1\right) = \frac{16}{3} \times \frac{3}{8} = 2$$

$$x = \frac{1}{3}(2 - 5 \times 3 + 16)$$

$$= \frac{1}{3} \times 3 = 1$$

$$x = 1;\ y = 2;\ z = 3$$ **Ans.**

Example 3.36. *Solve the Gaussion elimination method*

$$10x - 7y + 3z + 5w = 6$$
$$-6x + 8y - z - 4w = 5$$
$$3x + y + 4z + 11w = 2$$
$$5x - 9y - 2z + 4w = 7$$

Solution: Given equations are:

$$10x - 7y + 3z + 5w = 6$$
$$-6x + 8y - z - 4w = 5$$
$$3x + y + 4z + 11w = 2$$
$$5x - 9y - 2z + 4w = 7$$

The augmented coefficient matrix is:

$$\left[\begin{array}{cccc|c} 10 & -7 & 3 & 5 & 6 \\ -6 & 8 & -1 & -4 & 5 \\ 3 & 1 & 4 & 11 & 2 \\ 5 & -9 & -2 & 4 & 7 \end{array}\right]$$

By making $R_3 \to R_1$; $R_4 \to R_2$; $R_2 \to R_3$ and $R_1 \to R_4$, we get

$$\begin{bmatrix} 3 & 1 & 4 & 11 & 2 \\ 5 & -9 & -2 & 4 & 7 \\ -6 & 8 & -1 & -4 & 5 \\ 10 & -7 & 3 & 5 & 6 \end{bmatrix}$$

By replacing $R_2 \to \left(R_2 - \frac{5}{3}R_1\right)$; $R_3 \to (R_3 + 2R_1)$ and $R_4 \to \left(R_4 - \frac{10}{3}R_1\right)$

we get

$$\begin{bmatrix} 3 & 1 & 4 & 11 & 2 \\ 0 & -\frac{32}{3} & -\frac{26}{3} & -\frac{43}{3} & \frac{11}{3} \\ 0 & 10 & 7 & 18 & 9 \\ 0 & -\frac{31}{3} & -\frac{31}{3} & -\frac{95}{3} & -\frac{2}{3} \end{bmatrix}$$

By replacing $R_2 \to 3R_2$ and $R_4 \to 3R_4$, we get

$$\begin{bmatrix} 3 & 1 & 4 & 11 & 2 \\ 0 & -32 & -26 & -43 & 11 \\ 0 & 10 & 7 & 18 & 9 \\ 0 & -31 & -31 & -95 & -2 \end{bmatrix}$$

By replacing $R_2 \to (R_2 - R_4)$, we get

$$\begin{bmatrix} 3 & 1 & 4 & 11 & 2 \\ 0 & -1 & 5 & 52 & 13 \\ 0 & 10 & 7 & 18 & 9 \\ 0 & -31 & -31 & -95 & -2 \end{bmatrix}$$

By replacing $R_3 \to (R_3 + 10R_2)$ and $R_4 \to (-31R_2 + R_4)$, we get

$$\begin{bmatrix} 3 & 1 & 4 & 11 & 2 \\ 0 & -1 & 5 & 52 & 13 \\ 0 & 0 & 57 & 538 & 139 \\ 0 & 0 & -186 & -1707 & -405 \end{bmatrix}$$

By replacing $R_4 \to \frac{1}{3}R_4$, we get

$$\begin{bmatrix} 3 & 1 & 4 & 11 & 2 \\ 0 & -1 & 5 & 52 & 13 \\ 0 & 0 & 57 & 538 & 139 \\ 0 & 0 & -62 & -569 & -135 \end{bmatrix}$$

By replacing $R_3 \to (R_3 + R_4)$, we get

$$\begin{bmatrix} 3 & 1 & 4 & 11 & 2 \\ 0 & -1 & 5 & 52 & 13 \\ 0 & 0 & -5 & -31 & 4 \\ 0 & 0 & -62 & -569 & -135 \end{bmatrix}$$

By replacing $R_4 \to \left(R_4 - \frac{62}{5}R_3\right)$, we get

$$\begin{bmatrix} 3 & 1 & 4 & 11 & 2 \\ 0 & -1 & 5 & 52 & 13 \\ 0 & 0 & -5 & -31 & 4 \\ 0 & 0 & 0 & -185 & -185 \end{bmatrix}$$

Now, the resultant augmented matrix may be retranslated as

$$\begin{bmatrix} 3 & 1 & 4 & 11 \\ 0 & -1 & 5 & 52 \\ 0 & 0 & -5 & -31 \\ 0 & 0 & 0 & -185 \end{bmatrix} \begin{bmatrix} x \\ y \\ z \\ w \end{bmatrix} \begin{bmatrix} 2 \\ 3 \\ 4 \\ -185 \end{bmatrix}$$

By equating the fourth line, we get

$$-185w = -185$$

or $$w = 1$$

From 3rd row, we get

$$-5z - 31w = 4$$

or $$-5z - 31 = 4$$

or $$z = \frac{4+31}{-5} = -7$$

From 2nd row, we get

or $$-y + 5z + 52w = 3$$

$$-y + 5(-7) + 52 \times 1 = 13$$

$$-y - 35 + 52 = 13$$

$\therefore$ $$y = -35 + 52 - 13 = 4$$

From the first row, we get

$$3x + y + 4z + 11w = 2$$

$$3x + 4 + 4 \times (-7) + 11 \times 1 = 2$$

$$3x + 4 - 28 + 11 = 2$$

i.e. $$3x = 2 + 28 - 15 = 5$$

$$x = \frac{1}{3} \times 15 = 5.$$

$\therefore$ $x = 5; y = 4; z = -7; w = 1$ **Ans.**

Example 3.37. *Solve the following simultaneous equations by any suitable method.*

$$x + y + z = 9$$
$$2x - 3y + 4z = 13$$
$$3x + 4y + 5z = 40$$

Solution: Given equations are:

$$x + y + z = 9 \quad \ldots(1)$$
$$2x - 3y + 4z = 13 \quad \ldots(2)$$
$$3x + 4y + 5z = 40 \quad \ldots(3)$$

Replace eqn. (2) by eqn. (2) – 2 eqn. (1) and eqn. (3) by eqn. (3) – 3 eqn. (1)

$$x + y + z = 9 \quad \ldots(4)$$
$$-5y + 2z = -5 \quad \ldots(5)$$
$$y + 2z = 13 \quad \ldots(6)$$

Replace eqn. (4) by [eqn. (4) + $\frac{1}{5}$ eqn. (5)] and eqn. (6) by [eqn. (6) + $\frac{1}{5}$ eqn. (5)] to eliminate Y from eqns. (4) and (6)

$$x + \frac{7}{5}z = 8 \quad \ldots(7)$$
$$-5y + 2z = -5 \quad \ldots(8)$$
$$\frac{12}{5}z = 12 \quad \ldots(9)$$

Replace eqn. (7) by [eqn. (7) – $\frac{7}{12}$ eqn. (9)] and eqn. (8) by [eqn. (8) – $\frac{5}{8}$ eqn. (9)]

$$x = 1$$
$$-5y = -15$$
$$\frac{12}{5}z = 12$$

Hence, $x = 1$, $y = 3$, $z = 5$ **Ans.**

❑❑❑

4 Eigen Values and Eigen Vectors

4.1 INTRODUCTION

Eigen values and eigen vectors from the enginearing point of views are very important in connection with matrices and a number of research papers that involve numerical methods for computers.

Characteristic Matrix: The matrix obtained by subtracting the product of a scalar quantity and the unit matrix from any matrix, is called the *characteristic matrix* of the given matrix.

Let $A = \begin{bmatrix} 3 & 2 & 1 \\ 1 & 2 & 3 \\ 1 & 3 & 2 \end{bmatrix}$ and λ a scalar quantity.

then $$A - \lambda I = \begin{bmatrix} 3 & 2 & 1 \\ 1 & 2 & 3 \\ 1 & 3 & 2 \end{bmatrix} - \lambda \begin{bmatrix} 1 & 0 & 0 \\ 0 & 1 & 0 \\ 0 & 0 & 1 \end{bmatrix} = \begin{bmatrix} 3-\lambda & 2 & 1 \\ 1 & 2-\lambda & 3 \\ 1 & 3 & 2-\lambda \end{bmatrix}$$

$\therefore$ $\begin{bmatrix} 3-\lambda & 2 & 1 \\ 1 & 2-\lambda & 3 \\ 1 & 3 & 2-\lambda \end{bmatrix}$ is the characteristic matrix.

The polynomical obtained by expanding the determinant $(A - \lambda I)$, is called the *characteristic polynomial* of the matrix A.

$$\begin{vmatrix} 3-\lambda & 2 & 1 \\ 1 & 2-\lambda & 3 \\ 1 & 3 & 2-\lambda \end{vmatrix}$$

$= (3 - \lambda) \{(2 - \lambda)(2 - \lambda) - 9\} - 2\{1(2 - \lambda) - 3\} + 1\{1 \times 3 - (2 - \lambda)\}$

$= (3 - \lambda)(\lambda^2 - 4\lambda - 5) + 2\lambda + 2 + 1 + \lambda$

$= 3\lambda^2 - 12\lambda - 15 - \lambda^3 + 4\lambda^2 + 5\lambda + 2\lambda + 2 + 1 + \lambda$

$= -\lambda^3 + 7\lambda^2 - 4\lambda - 12$

is the required characteristic polynomial of the matrix A.

(*iii*) The equation $\{A - \lambda I\} = 0$ is called the characteristic equation of the matrix A. *i.e.,*

$\lambda^3 - 7\lambda^2 + 4\lambda + 12 = 0$ is the required charcteristic equation of the matrix A.

(*iv*) The roots of the characteristic equation $|A - \lambda I| = 0$ are called the characteristic roots or eigen values of the matrix.

Solution of the characteristic equation

$$\lambda^3 - 7\lambda^2 + 4\lambda + 12 = 0 \qquad ...(i)$$

Putting $\lambda = 2$ in eqn. (*i*), we get

$$8 - 28 + 8 + 12 = 0$$

$\therefore$ $(\lambda - 2)$ is one of the factors of the equation.

Dividing eqn. (*i*) by $(\lambda - 2)$, we get $\lambda^2 - 5\lambda - 6$

$$\therefore \quad \lambda^3 + 7\lambda^2 + 4\lambda + 12 = 0$$

$$(\lambda - 2)\ (\lambda - 6)\ (\lambda + 1) = 0$$

or $$\lambda = -1, 2, 6.$$

$\therefore$ Eigen values of the matrix are: $-1, 2, 6$

(*v*) **Eigen Vectors.** Corresponding to each eigen value of a matrix a corresponding non-zero vector x exists satisfying the equation $|A - \lambda I| = 0$

The non-zero vectors are called the *eigen vectors.*

Hence,

$$\begin{bmatrix} 3+1 & 2 & 1 \\ 1 & 2+1 & 3 \\ 1 & 3 & 2+1 \end{bmatrix} \begin{bmatrix} x_1 \\ x_2 \\ x_3 \end{bmatrix} =$$ 0 is the eigen vector for the eigen value $\lambda = -1$.

Example 4.1. *Find the eign values of the matrix*

$$\begin{vmatrix} 3 & 2 & 5 \\ 6 & -5 & 3 \\ -24 & 38 & 2 \end{vmatrix}$$

Solution: Characteristic matrix of the matrix

$$= A - \lambda I = \begin{bmatrix} 3 & 2 & 5 \\ 6 & -5 & 3 \\ -24 & 38 & 2 \end{bmatrix} - \lambda \begin{bmatrix} 1 & 0 & 0 \\ 0 & 1 & 0 \\ 0 & 0 & 1 \end{bmatrix}$$

$$= \begin{bmatrix} 3 & 2 & 5 \\ 6 & -5 & 3 \\ -24 & 38 & 2 \end{bmatrix} - \begin{bmatrix} \lambda & 0 & 0 \\ 0 & \lambda & 0 \\ 0 & 0 & \lambda \end{bmatrix}$$

$$= \begin{bmatrix} (3-\lambda) & 2 & 5 \\ 6 & -(5+\lambda) & 3 \\ -24 & 38 & (2-\lambda) \end{bmatrix}$$

$$= (3 - \lambda)\,[-(5 + \lambda)(2 - \lambda) - 114] - 2\,[6\,(2 - \lambda) + 72] + 5\,[228 - 24\,(5 + \lambda)]$$

$$= (3 - \lambda)\,[-(10 + 2\lambda - 5\lambda - \lambda^2 - 114)] - 2\,[12 - 6\lambda + 72] + 5\,(228 - 120 - 24\lambda)$$

$$= (3 - \lambda)\,(-(10 + 3\lambda + \lambda^2 - 114) - (24 - 12\lambda + 144) + 5\,(108 - 24\lambda)$$

$$= (3 - \lambda)\,(\lambda^2 + 3\lambda - 124) - (168 - 12\lambda) + 5\,(108 - 24\lambda)$$

$$= 3\lambda^2 + 9\lambda - 372 - \lambda^2 - 3\lambda^2 + 124\lambda - 168 + 12\lambda + 540 - 120\lambda$$

or $$-\lambda^3 + 25\lambda = 0$$

$$\lambda^3 - 25\lambda = 0$$

$$\lambda(\lambda^2 - 25) = 0$$

$$\lambda(\lambda + 5)(\lambda - 5) \Rightarrow \lambda = 0, -5, 5$$

∴ Hence, the equivalues of the matrix are

$$\lambda = 0, -5, 5 \text{ **Ans.**}$$

Example 4.2. *Find the characteristic equation of the matrix.*

$$P = \begin{bmatrix} 6 & -2 & 2 \\ -2 & 3 & -1 \\ 2 & -1 & 3 \end{bmatrix}$$

Also, find eign values and eigen vectors.

Solution: The characteristic equation is:

$$|P - \lambda I| = \begin{bmatrix} 3-\lambda & -2 & 2 \\ -2 & 3-\lambda & -1 \\ 2 & -1 & 3-\lambda \end{bmatrix} = 0$$

or $$(6 - \lambda)\,[(3 - \lambda)^2 - 1] + 2\,[-2\,(3 - \lambda) + 2] + 2\,[2 - 2\,(3 - \lambda)] = 0$$

$$(6 - \lambda)\,[9 - 6\lambda + \lambda^2) - 1] + 2\,[-6 + 2\lambda + 2\,[2 - 6 + 2\lambda] = 0$$

$$(6 - \lambda\,(8 - 6\lambda + \lambda^2) + 2\,(2\lambda - 4) + 2\,(2\lambda - 4)$$

$$48 - 36\lambda + 6\lambda^2 - 8\lambda + 6\lambda^2 - \lambda^3 + 4\lambda - 8 + 4\lambda - 8$$

$$-48 + 36\lambda - 6\lambda^2 + 8\lambda - 6\lambda^2 + \lambda^3 - 4\lambda + 8 - 4\lambda + 8 = 0$$

or $$\lambda^3 - 12\lambda^2 + 36\lambda - 32 = 0$$

or $$(\lambda - 2)(\lambda - 2)(\lambda - 8) = 0$$

∴ Eigen values are : 2, 2, 8

Corresponding to $\lambda = 2$, the eigen vectors are given by

$$(P - 2I)x = 0$$

or
$$\begin{bmatrix} 4 & -2 & 2 \\ -2 & 1 & -1 \\ 2 & -1 & 1 \end{bmatrix} \begin{bmatrix} x_1 \\ x_2 \\ x_3 \end{bmatrix} = 0 \quad \text{...}(i)$$

From equation (*i*), we get only one independent equation

$$2x_1 - x_2 + x_3 = 0$$

Choosing $x_2 = 0$, we get

$$2x_1 = -x_3$$

$$\frac{x_1}{1} = -\frac{x_3}{2}$$

$$\therefore \quad \frac{x_1}{1} = \frac{x_2}{0} = \frac{x_3}{-2} \text{ gives the eigen vector } (1, 0, -2)$$

Next choosing $x_3 = 0$,

$$2x_1 - x_2 = 0$$

$$x_1 = \frac{x_2}{2}$$

$$\therefore \quad \frac{x_1}{1} = \frac{x_2}{2} = \frac{x_3}{6}$$

gives another independent eigen vector (1, 2, 0)

Any other eigen vector corresponding to $\lambda = 2$ will be a linear combination of these two.

Now, corresponding to $\lambda = 9$, we get equation

$$-2x_1 - 2x_2 + 2x_3 = 0 \quad \text{...}(ii)$$

$$-2x = -5x_2 - x_3 = 0 \quad \text{...}(iii)$$

Solving equations (*ii*) and (*iii*), we get

$$\frac{x_1}{2} = \frac{x_2}{-1} = \frac{x_3}{1}\text{, to give an eigen vector } (2, -1, 1). \textbf{ Ans.}$$

Example 4.3. *Find the eigen values of the matrix A given by*

$$A = \begin{bmatrix} 2 & 6 \\ 1 & 1 \end{bmatrix}$$

Solution:

$$\text{A} = \begin{bmatrix} 2 & 6 \\ 1 & 1 \end{bmatrix}$$

∴ Characteristic matrix of the matrix A

$$A - \lambda I = \begin{bmatrix} 2 & 6 \\ 1 & 1 \end{bmatrix} - \lambda \begin{bmatrix} 1 & 0 \\ 0 & 1 \end{bmatrix} = \begin{bmatrix} 2-\lambda & 6 \\ 1 & 1-\lambda \end{bmatrix}$$

Characteristic polynomial of matrix A

$$= \begin{bmatrix} 2-\lambda & 6 \\ 1 & 1-\lambda \end{bmatrix} = (2 - \lambda)(1 - \lambda) - 6 \times 1$$

$$= \lambda^2 - 3\lambda - 4$$

Characteristic equation

$$= |A - \lambda I| = 0$$

$\therefore \quad \lambda^2 - 3\lambda - 4 = 0$ or $\lambda = 4$ or -1

$\therefore$ Eigen values are 4, – 1 **Ans.**

Example 4.4. *For the matrix.*

$$A = \begin{bmatrix} 1 & 0 & 2 \\ 2 & 1 & 3 \\ 2 & 0 & 3 \end{bmatrix}$$

Obtain all the eigen values of A from the definition of eigen value.

Solution:

Let $\quad A = \begin{bmatrix} 1 & 0 & 2 \\ 2 & 1 & 3 \\ 2 & 0 & 3 \end{bmatrix}$

Characteristic matrix

$$A - \lambda I = \begin{bmatrix} 1 & 0 & 2 \\ 2 & 1 & 3 \\ 2 & 0 & 3 \end{bmatrix} - \lambda \begin{bmatrix} 1 & 0 & 0 \\ 0 & 1 & 0 \\ 0 & 0 & 1 \end{bmatrix}$$

$$= \begin{bmatrix} 1-\lambda & 0 & 2 \\ 2 & 1-\lambda & 3 \\ 2 & 0 & 3-\lambda \end{bmatrix}$$

Characteristic polynomial of the matrix A

$$= \begin{bmatrix} 1-\lambda & 0 & 2 \\ 2 & 1-\lambda & 3 \\ 2 & 0 & 3-\lambda \end{bmatrix}$$

$= (1 - \lambda)(3 - 3\lambda - \lambda + \lambda^2) + 2(2 \times 0 - 2 + 2\lambda)$

$= (1 - \lambda)(3 - 4\lambda + \lambda^2) + 2(-2 + 2\lambda)$

$= 3 - 4\lambda + \lambda^2 - 3\lambda + 4\lambda^2 - \lambda^3 - 4 + 4\lambda$

$= -\lambda^3 + 5\lambda^2 - 3\lambda - 1$

Characteristic equation $-\lambda^3 + 5\lambda^2 - 3\lambda - 1 = 0$

or $\lambda^3 - 5\lambda^2 + 3\lambda + 1 = 0$

∴ Eigen values of the matrix $\lambda = 1, \lambda = 2 \pm \sqrt{5}$ **Ans.**

Example 4.5. *Find the eigen values of the matrix.*

$$\begin{bmatrix} 8 & -6 & 2 \\ -6 & 7 & -4 \\ 2 & -4 & 3 \end{bmatrix}$$

Solution:

Characteristic matrix $|A - \lambda I|$

$$= \begin{bmatrix} 8 & -6 & 2 \\ -6 & 7 & -4 \\ 2 & -4 & 3 \end{bmatrix} - \lambda \begin{bmatrix} 1 & 0 & 0 \\ 0 & 1 & 0 \\ 0 & 0 & 1 \end{bmatrix}$$

$$= \begin{bmatrix} 8 & -6 & 2 \\ -6 & 7 & -4 \\ 2 & -4 & 3 \end{bmatrix} - \begin{bmatrix} \lambda & 0 & 0 \\ 0 & \lambda & 0 \\ 0 & 0 & \lambda \end{bmatrix}$$

$$= \begin{bmatrix} (8-\lambda) & -6 & 2 \\ -6 & (7-\lambda) & -4 \\ 2 & -4 & (3-\lambda) \end{bmatrix}$$

$= (8-\lambda)\{(7-\lambda)(3-\lambda) - 16\} - (-6)\{-6(3-\lambda) + 8\} + 2\{24 - 2(7-\lambda)\}$

$= (8-\lambda)\{21 - 3\lambda - 7\lambda + \lambda^2 - 16\} + 6\{-18 + 6\lambda + 8\} + 2\{24 - 14 + 2\lambda\}$

$= (8-\lambda)(\lambda^2 - 10\lambda + 5) + 6(6\lambda - 10) + 2(2\lambda + 10)$

$= -\lambda^3 + 18\lambda^2 - 85\lambda + 40 + 36\lambda - 60 + 4\lambda + 20$

$-\lambda^3 + 18\lambda^2 - 45\lambda = 0$

$\lambda^3 - 18\lambda^2 + 45\lambda = 0$

$\lambda(\lambda^2 - 18\lambda + 45) = 0$

i.e., $\lambda = 0, \lambda = 3, \lambda = 15$

or Eigen values are : 0, 3, 15 **Ans.**

Example 4.6. *Find the characteristic values and characteristic vectors of matrix*

$$A = \begin{bmatrix} 2 & 1 & 2 \\ 1 & 2 & 2 \\ 2 & 2 & 1 \end{bmatrix}$$

Solution: The Characteristic matrix

$$A - \lambda I = \begin{bmatrix} 2 & 1 & 2 \\ 1 & 2 & 2 \\ 2 & 2 & 1 \end{bmatrix} - \lambda \begin{bmatrix} 1 & 0 & 0 \\ 0 & 1 & 0 \\ 0 & 0 & 1 \end{bmatrix} \begin{bmatrix} 2-\lambda & 1 & 2 \\ 1 & 2-\lambda & 2 \\ 2 & 2 & 1-\lambda \end{bmatrix}$$

The Characteristic matrix is

$$\begin{bmatrix} 2-\lambda & 1 & 2 \\ 1 & 2-\lambda & 2 \\ 2 & 2 & 1-\lambda \end{bmatrix} = 0$$

or $(2 - \lambda) [(2 - \lambda) (1 - \lambda) - 2 \times 2] - 1 [(1 - \lambda) - 2 \times 2] + 2 [2 \times 1 - 2 (2 - \lambda] = 0$

or $(2 - \lambda) (\lambda^2 - 3 \lambda + 2 - 4) - (1 - \lambda - 4) + 2 (2 - 4 + 2 \lambda) = 0$

or $(2 - \lambda) (\lambda^2 - 3 \lambda - 2) - (- \lambda - 3) + 2 (- 2 + 2 \lambda) = 0$

or $- \lambda^3 + 5 \lambda^2 - 4 \lambda - 4 + \lambda + 3 - 4 + 4 \lambda = 0$

or $- \lambda^3 + 5 \lambda^2 + \lambda - 5 = 0$

or $\lambda^3 - 5 \lambda^2 - \lambda + 5 = 0$

$(\lambda^2 - 1) (\lambda - 5) = 0$

$\therefore \quad \lambda = 1, -1, 5$

$\therefore$ The characteristic values of matrix A are :

$\lambda_1 = 1, \lambda_2 = -1, \lambda = 5$ **Ans.**

The characteristic vector X has to satisfy the equation

$[A - \lambda I] X = 0$ are the equations for the given matrix A,

$$\begin{bmatrix} 2-\lambda & 1 & 2 \\ 1 & 2-\lambda & 2 \\ 2 & 2 & 1-\lambda \end{bmatrix} \begin{bmatrix} x_1 \\ x_2 \\ x_3 \end{bmatrix} = \begin{bmatrix} 0 \\ 0 \\ 0 \end{bmatrix} \qquad ...(i)$$

Corresponding to $\lambda_1 = 1$, the characteristic vector X_1 is :

$$\begin{bmatrix} 1 & 1 & 2 \\ 1 & 1 & 2 \\ 2 & 2 & 0 \end{bmatrix} \begin{bmatrix} x_1 \\ x_2 \\ x_3 \end{bmatrix} = \begin{bmatrix} 0 \\ 0 \\ 0 \end{bmatrix}$$

i.e.,

$$\left.\begin{aligned} x_1 + x_2 + x_3 &= 0 \\ x_1 + x_2 + 2 x_3 &= 0 \\ 2 x_1 + 2 x_2 + 0 &= 0 \end{aligned}\right\} \qquad ...(ii)$$

Solving the first two equations of (*ii*), we get proportional values of x_1, x_2 and x_3 *i.e.,*

$$\frac{x_1}{2-1} = \frac{x_2}{1-2} = \frac{x_3}{1-1} \text{ i.e., } \frac{x_1}{1} = \frac{x_1}{-1} = \frac{x_3}{0} = k$$

For K = 1, the eigen vector corresponding to $\lambda_1 = 1$ is

$x_1 = 1, x_2 = -1, x_3 = 0$, *i.e.,*

$X_1 = [1, -1, 0]^T$

For $\lambda = -1$, the equations from eqn. (*i*) are :

$3 x_1 + x_2 + 3 x_3 = 0$

$$x_1 + 3x_2 + 2x_3 = 0 \qquad ...(iii)$$
$$2x_1 + 2x_2 + 2x_3 = 0$$

Solving the first two equations of eqn. (*iii*), we get proportional values of x_1, x_2 and x_3 *i.e.*,

$$\frac{x_1}{2-3} = \frac{x_2}{1-6} = \frac{x_3}{9-1} = k$$

For K = 1, the characteristic vector corresponding to $\lambda_2 = -1$,

$$X_2 = [-1, -5, 8]^T$$

Similarly, for $\lambda = 5$, the equations are

$$-3x_1 + x_2 + 2x_3 = 0$$
$$x_1 + 3x_2 + 2x_3 = 0 \qquad ...(iv)$$
$$2x_1 + 2x_2 + 4x_3 = 0$$

Solving the first two equations of eqn. (*iv*), we get proportional values of x_1, x_2 and x_3

$$\frac{x_1}{2+6} = \frac{x_2}{2+6} = \frac{x_3}{-9-1} = k$$

or
$$\frac{x_1}{8} = \frac{x_2}{8} = \frac{x_3}{-10} = k$$

For $k = 1$, the characteristic vector corresponding to $\lambda_3 = 5$.

$$x_3 = [8\ 8\ -10]^T$$

The three eigen vectors are :

Example. 4.7. *Compute the Eigen values and the corresponding Eigen vectors of the following systems :*

$$10x_1 + 2x_2 + x_3 = \lambda x_1$$
$$2x_1 + 10x_2 + x_3 = \lambda x_2$$
$$2x_1 + x_2 + 10x_3 = \lambda x_3$$

Solution: The given equation systems are

$$10x_1 + 2x_2 + x_3 = \lambda x_1$$
$$2x_1 + 10x_2 + x_3 = \lambda x_2$$
$$2x_1 + x_2 + 10x_3 = \lambda x_3$$

The corresponding Eigen-matrix is :

$$\begin{bmatrix} 10 & 2 & 1 \\ 2 & 10 & 1 \\ 2 & 1 & 10 \end{bmatrix} - \lambda \begin{bmatrix} 1 & 0 & 0 \\ 0 & 1 & 0 \\ 0 & 0 & 1 \end{bmatrix}$$

$$\begin{bmatrix} 10-\lambda & 2 & 1 \\ 2 & 10-\lambda & 1 \\ 2 & 1 & 10-\lambda \end{bmatrix} = 0$$

$\therefore$ Characteristic polynomial matrix may be obtained by expanding with the help of first column.

$= (10 - \lambda)[(10 - \lambda)(10 - \lambda) - 1] - 2[2(10 - \lambda) - 1] + 1[2.1 - 1(10 - \lambda)]$

$= (10 - \lambda)[\lambda^2 - 20\lambda + 99] - 2[20 - 2\lambda - 1] + 2[2 - 10 + \lambda]$

$= [10\lambda^2 - 200\lambda + 990 - \lambda^3 + 20\lambda^2 - 99\lambda] + [4\lambda - 38] + [2\lambda - 16]$

$= [-\lambda^3 + 30\lambda^2 - 299\lambda + 990 + 2\lambda - 38 + 2\lambda - 16]$

$= [\lambda^3 - 30\lambda^2 + 293\lambda - 936]$

$= \lambda^3 - 30\lambda^2 + 293\lambda - 936 = 0$

By trial we note that

$$8^3 - 30 \times 8^2 + 293 \times 8 - 936 = 0$$
$$512 - 1920 + 2344 - 936 = 0$$
$$2856 - 2856 = 0$$

$\therefore$ $\lambda = 8$ is one of the root.

Similarly, we get

$$9^3 - 30 \times 9^2 + 293 \times 9 - 936 = 0$$

$729 - 2430 + 2637 - 936\ 3366 - 3366 = 0$

$\therefore$ $\lambda = 9$ is also one of the root.

Similarly we get $\lambda = 11$ as the third root.

$\therefore$ Eigen values are : $\lambda_1 = 8, \lambda_2 - 9, \lambda_3 = 13$ **Ans.**

Eigen Vectors :

For $\lambda = 8$, we get

$$\begin{bmatrix} 2 & 2 & 1 \\ 2 & 2 & 1 \\ 2 & 1 & 2 \end{bmatrix} \begin{bmatrix} x_1 \\ x_2 \\ x_3 \end{bmatrix} = 0$$

$$2x_1 + 2x_2 + x_3 = 0 \quad \ldots(i)$$
$$2x_1 + 2x_2 + x_3 = 0 \quad \ldots(ii)$$
$$2x_1 + x_2 + 2x_3 = 0 \quad \ldots(iii)$$

Eliminating x_1 for equations (*ii*) and (*iii*), we get

$$+(2x_2 + x_3) = +(x_2 + 2x_3)$$
$$2x_2 - x_2 = 2x_3 - x_3$$
$$x_2 = x_3$$

or
$$\frac{x_2}{1} = \frac{x_3}{1}$$

From Equation (*i*), we get

$$2x_1 + 2x_2 + x_2 = 0$$
$$2x_1 + 3x_2 = 0$$

$$\therefore \qquad 2x_1 = -3x_2$$

$$\frac{x_1}{-3} = \frac{x_2}{2} = \frac{x_3}{2}$$

$\therefore$ The corresponding Eigen vector for $\lambda = 8$ is

$$[-3 \quad 2 \quad 2]$$

For $\lambda = 9$, we get

$$\begin{bmatrix} 1 & 2 & 1 \\ 2 & 1 & 1 \\ 2 & 1 & 2 \end{bmatrix} \begin{bmatrix} x_1 \\ x_2 \\ x_3 \end{bmatrix} = 0$$

$$x_1 + 2x_2 + x_3 = 0 \qquad \text{...}(v)$$

$$2x_1 + x_2 + x_3 = 0 \qquad \text{...}(vi)$$

$$2x_1 + x_2 + x_3 = 0 \qquad \text{...}(vii)$$

From equations (*v*) and (*vi*), we get

$$\frac{x_1}{2-1} = \frac{x_2}{2-1} = \frac{x_3}{1-4}$$

$$\frac{x_1}{1} = \frac{x_2}{1} = \frac{x_3}{-3}$$

$\therefore$ Second Eigen vector is :

$$[1 \quad 1 \quad -3]$$

Similarly for $\lambda = 11$, we get

$$\begin{bmatrix} -1 & 2 & 1 \\ 2 & -1 & 1 \\ 2 & 1 & -1 \end{bmatrix} \begin{bmatrix} x_1 \\ x_2 \\ x_3 \end{bmatrix} = 0$$

$$-x_1 + 2x_2 + x_3 = 0 \qquad \text{...}(viii)$$

$$2x_1 - x_2 + x_3 = 0 \qquad \text{...}(ix)$$

$$2x_1 + x_2 - x_3 = 0 \qquad \text{...}(x)$$

Eliminating x_1 for equations (*viii*) and (*ix*), we get

$$\frac{x_2 - x_3}{2} = \frac{x_3 - x_2}{2}$$

or $\qquad x_2 + x_2 = x_3 + x_3$ or $2x_2 = 2x_3$

or $$\frac{x_2}{1} = \frac{x_3}{1}$$

Substituting $x_3 = x_2$ in equation (*vii*), we get

$$-x_3 + 2x_2 + x_2 = 0$$

or

$$-x_1 = -3x_2$$

$$x_1 = 3x_2$$

$$\frac{x_1}{3} = \frac{x_2}{1} = \frac{x_3}{1}$$

∴ Second Eigen vector is:

[3 1 1] **Ans.**

4.2 PROPERTIES OF EIGEN VALUES

The following are the important of eigen values:

(*i*) The sum of the eigen values of a matrix A, is equal to the sum of the elements of its principal diagonal.

(*ii*) If λ is an eigen value of an orthogonal matrix, then I/λ is also its eigen value.

(*iii*) If λ is an eigen value of a matrix A, then I/λ is the eigen value of A^{-1}.

(*iv*) If $\lambda_1, \lambda_2, \lambda_3 ..., \lambda_n$ are the eigen values of a matrix A, then A^m has the eigen values $\lambda_1^m, \lambda_2^m, \lambda_3^m, ... \lambda_n^m$ where m is a positive integer.

(*v*) Any similarity transformation applied to a matrix leaves its eigen values uncharged.

(*vi*) If a square matrix A has n linearly independent eigen vectors, then a matrix P can be found such that $P^{-1}AP$ is a diagonal matrix whose diagonal elements are the eigen values of A.

(*vii*) The transformation of A by a non-singular matrix P to $P^{-1}AP$ is called a similarity transformation.

4.3 MATRIX INVERSION BY CAYLEY HAMILTON THEOREM

According to Cayley Hamilton Theorem, "*every square matrix satisfies its characteristic equation.*"

While finding matrix inversion by this method, the following steps are involved.

(*i*) Calculate the characteristic equation.

(*ii*) Replace λ by A to obtain the equation in A.

(*iiii*) Multiply the equation thus forned by A^{-1}.

(*iv*) Calculate the value of A^{-1}.

The method of obtaining the inverse may be studied by following the following solved example.

Example 4.8. *Obtain the inverse of the matrix*

$$A = \begin{bmatrix} 2 & 1 & 3 \\ 1 & 3 & 2 \\ 2 & 1 & 4 \end{bmatrix}$$

Solution: The characteristic equation is:

$$= \begin{bmatrix} 2 & 1 & 3 \\ 1 & 3 & 2 \\ 2 & 1 & 4 \end{bmatrix} - \lambda \begin{bmatrix} 1 & 0 & 0 \\ 0 & 1 & 0 \\ 0 & 0 & 1 \end{bmatrix} = 0$$

or
$$d\begin{bmatrix} 2-\lambda & 1 & 3 \\ 1 & 3-\lambda & 2 \\ 2 & 1 & 4-\lambda \end{bmatrix} = 0$$

$$(2-\lambda)\{(3-\lambda)(4-\lambda)-2\} - 1\{(4-\lambda)x1 - 4\} + 3\{1 \times 1 - 2(3-\lambda)\} = 0$$

$$(2-\lambda)(12-7\lambda+\lambda^2-2) - (4-\lambda-4) + 3(1-6+2\lambda)\} = 0$$

$$(2-\lambda)(\lambda^2-7\lambda+10) + \lambda + 6\lambda - 15 = 0$$

$$2\lambda^2 - 14\lambda + 20 - \lambda^3 + 7\lambda^2 - 10\lambda + \lambda + 6\lambda - 15 = 0$$

$$-\lambda^3 + 9\lambda^2 - 17\lambda + 5 = 0$$

$$\lambda^3 - 9\lambda^2 + 17\lambda - 5 = 0 \quad ...(i)$$

Substituting A for λ in eqn, (*i*), we get

$$A^3 - 9A^2 + 17A - 5 = 0 \quad ...(ii)$$

Multiplying the eqn. (*ii*) by A^{-1}, we get

$$A^2 - 9A + 17I - 5A^{-1} = 0$$

$$\therefore \quad 5A^{-1} = A^2 + 9A + 17I$$

$$= \begin{bmatrix} 2 & 1 & 3 \\ 1 & 3 & 2 \\ 2 & 1 & 4 \end{bmatrix}\begin{bmatrix} 2 & 1 & 3 \\ 1 & 3 & 2 \\ 2 & 1 & 4 \end{bmatrix} - 9\begin{bmatrix} 2 & 1 & 3 \\ 1 & 3 & 2 \\ 2 & 1 & 4 \end{bmatrix} + 17\begin{bmatrix} 1 & 0 & 0 \\ 0 & 1 & 0 \\ 0 & 0 & 1 \end{bmatrix}$$

$$= \begin{bmatrix} 11 & 8 & 20 \\ 9 & 12 & 17 \\ 13 & 9 & 24 \end{bmatrix} - \begin{bmatrix} 18 & 9 & 27 \\ 9 & 27 & 18 \\ 18 & 9 & 36 \end{bmatrix} + \begin{bmatrix} 17 & 0 & 0 \\ 0 & 17 & 0 \\ 0 & 0 & 17 \end{bmatrix}$$

$$= \begin{bmatrix} 10 & -1 & -7 \\ 0 & 2 & -1 \\ 5 & 0 & +5 \end{bmatrix}$$

$$\therefore \quad A^{-1} = \frac{1}{5}\begin{bmatrix} 10 & -1 & -7 \\ 0 & 2 & -1 \\ 5 & 0 & 5 \end{bmatrix} = \begin{bmatrix} 0 & -1/5 & -7/5 \\ 0 & 2/5 & -1/6 \\ -1 & 0 & +1 \end{bmatrix}$$

Example 4.9. *Obtain the inverse of the matrix*

$$A = \begin{bmatrix} 8 & -8 & -2 \\ 4 & -3 & -2 \\ 3 & -4 & 1 \end{bmatrix}$$

Solution: The characteristic equation of the matrix A

$$\begin{bmatrix} 8-\lambda & -8 & -2 \\ 4 & -3-\lambda & -2 \\ 3 & -4 & 1-\lambda \end{bmatrix} = 0$$

or $(8-\lambda)\,[-3-\lambda)\,(1-\lambda)-8] + 8\,[4(1-\lambda)+6] - 2[-16-3(-3-\lambda)]$

or $(8-\lambda)\,[\lambda^2+2\lambda-11] + 8(10-4\lambda) - 2(3\lambda-7)$

or $8\lambda^2 + 16\lambda - 88 - \lambda^3 - 2\lambda^2 + 11\lambda + 80 - 32\lambda - 6\lambda + 14$

or $-\lambda^3 + 6\lambda^2 - 11\lambda + 6 = 0$

$\lambda^3 - 6\lambda^2 + 11\lambda - 6 = 0$

Applying Cayley-Hemilton theorem, we get

$A^3 - 6A^2 + 11A - 6\,I = 0$ ...(*i*)

Multiplying the eqn. (*i*) by A^{-1}, we get

$A^2 - 6A + 11I - 6A^{-1} = 0$

or $6A^{-1} = A^2 - 6A + 11I$

$$= \begin{bmatrix} 8 & -8 & -2 \\ 4 & -3 & -2 \\ 3 & -4 & 1 \end{bmatrix}\begin{bmatrix} 8 & -8 & -2 \\ 4 & -3 & -2 \\ 3 & -4 & 1 \end{bmatrix} - 6\begin{bmatrix} 8 & -8 & -2 \\ 4 & -3 & -2 \\ 3 & -4 & 1 \end{bmatrix} + 11\begin{bmatrix} 1 & 0 & 0 \\ 0 & 1 & 0 \\ 0 & 0 & 1 \end{bmatrix}$$

or $$6A^{-1} = \begin{bmatrix} 26 & -32 & -2 \\ 14 & -15 & -4 \\ 11 & -16 & 3 \end{bmatrix} - \begin{bmatrix} 48 & -48 & -12 \\ 24 & -18 & -12 \\ 18 & -24 & 6 \end{bmatrix} + \begin{bmatrix} 11 & 0 & 0 \\ 0 & 11 & 0 \\ 0 & 0 & 11 \end{bmatrix}$$

$$= \begin{bmatrix} 26-48+11 & -32+48+0 & -2+12 & 0 \\ 14-24+0 & -15+18+11 & -4+12 & 0 \\ 11-18 & 0-16+24 & 0+3 & -6+11 \end{bmatrix}$$

$$= \begin{bmatrix} -11 & 16 & 10 \\ -10 & 14 & 8 \\ -7 & 8 & 8 \end{bmatrix}$$

$$\therefore \quad A^{-1} = \frac{1}{6}\begin{bmatrix} -11 & 16 & 10 \\ -10 & 14 & 8 \\ -7 & 8 & 8 \end{bmatrix} \quad \textbf{Ans.}$$

Check: $$AA^{-1} = I\begin{bmatrix} 8 & -8 & -2 \\ 4 & -3 & -2 \\ 3 & -4 & 1 \end{bmatrix} \times \frac{1}{6}\begin{bmatrix} -11 & 16 & 10 \\ -10 & 14 & 8 \\ -7 & 8 & 8 \end{bmatrix}$$

$$= \frac{1}{6}\begin{bmatrix} -88+80+14 & 128-112-16 & 80-63-16 \\ -44+30+14 & 64-42-16 & 40-24-16 \\ -33+40-7 & 48-56+8 & 30-32+8 \end{bmatrix}$$

$$= \frac{1}{6}\begin{bmatrix} 6 & 0 & 0 \\ 0 & 6 & 0 \\ 0 & 0 & 6 \end{bmatrix} = \begin{bmatrix} 1 & 0 & 0 \\ 0 & 1 & 0 \\ 0 & 0 & 1 \end{bmatrix} = I$$

Example 4.10. *Verify Caley Hamilton theorem for the matrix*

$$A = \begin{bmatrix} 2 & -1 & 1 \\ -1 & 2 & -1 \\ 1 & -1 & 2 \end{bmatrix}$$

Also, find the value of A^{-1}.

Solution: The character equation is

$$|A - \lambda I| = \begin{bmatrix} 2-\lambda & -1 & 1 \\ -1 & 2-\lambda & -1 \\ 1 & -1 & 2-\lambda \end{bmatrix} = 0$$

$\therefore$ $(2-\lambda)[(2-\lambda)(2-\lambda)-1]-(-1)[-1\times(2-\lambda)+1][+1-(2-\lambda)]$

or $(2-\lambda)[4-4\lambda+\lambda^2-1]+[-2+\lambda+1]+[\lambda-1]$

or $(2-\lambda)[\lambda^2-4\lambda+3]+[\lambda-1]+[\lambda-1]$

$2\lambda^2-8\lambda+6-\lambda^3+4\lambda^3-3\lambda+\lambda-1+\lambda-1$

or $$-\lambda^3+6\lambda^2-9\lambda+4 = 0$$

or $$\lambda^3-6\lambda^2+9\lambda-4 = 0$$

Now, to verify the Cayley Hamilton theorem, we must prove that

$A^3-6A^2+9A-4I = 0$

$$A^2 = \begin{bmatrix} 2 & -1 & 1 \\ -1 & 2 & -1 \\ 1 & -1 & 2 \end{bmatrix} \times \begin{bmatrix} 2 & -1 & 1 \\ -1 & 2 & -1 \\ 1 & -1 & 2 \end{bmatrix}$$

$= 2\times2+(-1)\times(-1)+1\times1, 2\times(-1)+(-1)(2)+1(1), 2\times1+(-1)(-1)$
$+1(2)-1\times2+2\times(-1)+-1(1), -1\times(-1)+2\times2+(-1)(-1), -1\times1+$
$2\times(-1)+(-1)(2)\ 1\times2+(-1)(-1)+2\times1, 1\times(-1)+(-1)(2)+2(-1),$
$1\times1+(-1)(-1)+2\times2$

$$= \begin{bmatrix} 6 & -5 & 5 \\ -5 & 6 & -5 \\ 5 & -5 & 6 \end{bmatrix}$$

$$A^3 = A^2 \times A = \begin{bmatrix} 6 & -5 & 5 \\ -5 & 6 & -5 \\ 5 & -5 & 6 \end{bmatrix} \times \begin{bmatrix} 2 & -1 & 1 \\ -1 & 2 & -1 \\ 1 & -1 & 2 \end{bmatrix}$$

$$\begin{bmatrix} 6\times 2+(-5)5\times 1+5\times 1, & 6\times(-1)+(-5)(2)+5(-1), & 6\times 1+(-5)(-1)+5\times 2 \\ -5\times 2+6(-1)+(-5)(1), & -5(-1)+6\times 2+(-5)(-1), & (-5)\times 1+6(-1)+5\times 2 \\ 5\times 2+(-5)(-1)+6\times 1, & 5\times(-1)+(-5)(+2)+6(-1), & 5\times 1+(-5)(-1)+6\times 2 \end{bmatrix}$$

or $$A^3 = \begin{bmatrix} 22 & -22 & 21 \\ -21 & 22 & -21 \\ 21 & -21 & 22 \end{bmatrix}$$

$$\therefore A^3 - 6A^2 + 9A - 4I = 0$$

$$\begin{bmatrix} 22 & -22 & 21 \\ -21 & 22 & -21 \\ 21 & -21 & 22 \end{bmatrix} - 6\begin{bmatrix} 6 & -5 & 5 \\ -5 & 6 & -5 \\ 5 & -5 & 6 \end{bmatrix} + 9\begin{bmatrix} 2 & -1 & 1 \\ -1 & 2 & -1 \\ 1 & -1 & 2 \end{bmatrix} - 4\begin{bmatrix} 1 & 0 & 0 \\ 0 & 1 & 0 \\ 0 & 0 & 1 \end{bmatrix}$$

$$\begin{bmatrix} 22 & -22 & 21 \\ -21 & 22 & -21 \\ 21 & -21 & 22 \end{bmatrix} - \begin{bmatrix} 36 & 30 & 30 \\ -30 & 36 & -30 \\ 30 & -30 & 36 \end{bmatrix} + 9\begin{bmatrix} 18 & -9 & 9 \\ -9 & 18 & -9 \\ 9 & -8 & 18 \end{bmatrix} - 4\begin{bmatrix} 4 & 0 & 0 \\ 0 & 4 & 0 \\ 0 & 0 & 4 \end{bmatrix}$$

$$\begin{bmatrix} 0 & 0 & 0 \\ 0 & 0 & 0 \\ 0 & 0 & 0 \end{bmatrix} = 0$$

This verifies the theorem.

To obtain A^{-1} multiply both the sides of eqn. (*i*) by A^{-1}, *i.e.*,

$$A^2 - 6A + 9I - 4A^{-1} = 0$$

$$= \begin{bmatrix} 6 & -5 & 5 \\ -5 & 6 & -5 \\ 5 & -5 & 5 \end{bmatrix} - 6\begin{bmatrix} 2 & -1 & 1 \\ -1 & 2 & -1 \\ 1 & -1 & 2 \end{bmatrix} + 9\begin{bmatrix} 1 & 0 & 0 \\ 0 & 1 & 0 \\ 0 & 0 & 1 \end{bmatrix}$$

$$= \begin{bmatrix} 3 & 1 & -1 \\ 1 & 3 & 1 \\ -1 & 1 & 3 \end{bmatrix}$$

$$\therefore \qquad A^{-1} = \frac{1}{4}\begin{bmatrix} 3 & 1 & -1 \\ 1 & 3 & 1 \\ -1 & 1 & 3 \end{bmatrix} \quad \textbf{Ans.}$$

Check: $AA^{-1} = I$

$$= \begin{bmatrix} 2 & -1 & 1 \\ -1 & 2 & -1 \\ 1 & -1 & 2 \end{bmatrix} \times \frac{1}{4}\begin{bmatrix} 3 & 1 & -1 \\ 1 & 3 & 1 \\ -1 & 1 & 3 \end{bmatrix}$$

$$= \frac{1}{4}\begin{bmatrix} 2 & -1 & 1 \\ -1 & 2 & -1 \\ 1 & -1 & 2 \end{bmatrix} \times \begin{bmatrix} 3 & 1 & -1 \\ 1 & 3 & 1 \\ -1 & 1 & 3 \end{bmatrix} = \frac{1}{4}\begin{bmatrix} 4 & 0 & 0 \\ 0 & 4 & 0 \\ 0 & 0 & 4 \end{bmatrix}$$

$$= \frac{1}{4}\begin{bmatrix} 6-1-1 & 3+2+1 & 3-1-2 \\ 2-3+1 & -1+6-1 & 1-3+2 \\ -2-1+3 & +1+2-3 & -1-1+6 \end{bmatrix}$$

$$= \begin{bmatrix} 1 & 0 & 0 \\ 0 & 1 & 0 \\ 0 & 0 & 1 \end{bmatrix} = I$$

4.4 CAYLEY HAMILTON THEOREM

If the characteristic equation of an n-rowed square matrix A be

$$\lambda^n + a_1\lambda^{n-1} + a_2\lambda^{n-2} + \ldots + c_n = 0$$

then

$$A^n + a_1A^{n-1} + a_2A^{n-2} + \ldots + a_nI = 0$$

The above theorem can also be re-stated by saying that *"every square matrix satisfies its characteristic equation."*

Proof. Since the elements of $(A - \lambda I)$ are at most of the first degree in λ, the elements of *adj* $(A - \lambda I)$ are atmost of degree $(n - 1)$ in λ.

$\therefore$ *adj* $(A - \lambda I)$ may be written as a matrix polynomial in λ

$$adj(A - \lambda I) = B_e\lambda^{n-1} + B_1{}^{n-2} + \ldots + B_{n-1} \qquad \ldots(1)$$

where B_0, B_1, B_{n-1} are $n \times n$ matrices, the elements being the functions of *adj*.

Again let

$$|A - \lambda I| = (-1)^n[\lambda^n + a_1\lambda^{n-1} + a_2\lambda^{n-2} + a_n] \qquad \ldots(2)$$

But

$$(A - \lambda I)\, adj(A - \lambda I) = |A - \lambda I| \qquad ...(3)$$

$$[\because \quad A\, adj\, A = (adj\, A)\, A = |A|\, I]$$

Using (1) and (2) in (3), we get

$$(A - \lambda I)\, B_0\lambda^{n-3} + B_1 l^{n-2} + ... + B_{n-1})$$
$$= (-1)^n [l^n + a_1\lambda^{n-1} + ... + a_n]I$$

Comparing co-efficients of like powers of λ on both sides, we have

$$-IB_0 = (-1)^n I$$
$$AB_0 - IB_1 = (-1)^n a_1 I$$
$$AB_1 - IB_2 = (-1)^n a_2 I$$
$$..............................$$
$$..............................$$
$$..............................$$
$$AB_{n-1} = (-1)^n A_n I$$

Pre-multiplying the above equations by A^n, A^{n-1} ..., I respectively and adding, we have

$$0 = (-1)^n [A^n + a_1 A^{n-1} + a_2 A^{n-2} + ... + a_n I]$$

Thus

$$A^n + a_1 A^{n-1} + ... + a_2 I = 0 \qquad ...(4)$$

4.5 CALCULATION OF INVERSE OF NON-SINGULAR MATRIX

If A be non singular, so that A^{-1} exists. By then pre-multiplying eqn. (4) throughout by A^{-1}, we have

$$A^{n-1} + a^1 A^{n-2} + a_2 A^{n-3} + ... + a_{n-1} I + a_{n-1} I + a_n A^{-1} = 0$$

showing that A^{-1} is expressible in terms of IA, A^2, ... A^{n-1}.

Example 4.11. *Obtain the characteristic equation of the matrix*

$\begin{pmatrix} 1 & 0 & 2 \\ 0 & 1 & 2 \\ 1 & 2 & 0 \end{pmatrix}$ *and hence calculate its inverse.*

Solution: Let $A = \begin{pmatrix} 1 & 0 & 2 \\ 0 & 1 & 2 \\ 1 & 2 & 0 \end{pmatrix}$

then $A - \lambda I = \begin{pmatrix} 1 & 0 & 2 \\ 0 & 1 & 2 \\ 1 & 2 & 0 \end{pmatrix} - \lambda \begin{pmatrix} 1 & 0 & 0 \\ 0 & 1 & 0 \\ 0 & 0 & 1 \end{pmatrix}$

$$= \begin{pmatrix} 1-\lambda & 0 & 2 \\ 0 & 1-\lambda & 2 \\ 1 & 2 & -\lambda \end{pmatrix}$$

$$\therefore \qquad |A - I\lambda| = \begin{pmatrix} 1-\lambda & 0 & 2 \\ 0 & 1-\lambda & 2 \\ 1 & 2 & -\lambda \end{pmatrix}$$

$$= (1-\lambda)[-\lambda(1-\lambda)-4] + 2[-(1-\lambda)]$$
$$= (1-\lambda)[\lambda^2-\lambda-4] - 2[1-\lambda]$$
$$= (1-\lambda)[\lambda^2-\lambda-6] = \lambda^2-\lambda-6-\lambda^3+\lambda^2+6\lambda$$
$$= -\lambda^3+2\lambda^2+5\lambda-6,$$

The characteristic equation of A

$$[A - \lambda I] = 0$$

i.e.,
$$\lambda^3 - 2\lambda^2 - 5\lambda + 6 = 0$$

To calculate the inverse of matrix A, we use Cayley-Hamilton theorem, *viz, every square matrix satisfies its characteristic equation.'*

$$A^3 - 2A^2 - 5A + 6I = 0 \qquad ...(1)$$

Multiplying (1) throughout by A^{-1}, we have

$$A^2 - 2A - 5I + 6A^{-1} = 0$$

or
$$-6A^{-1} = A^2 - 2A - 5I$$

$$= \begin{pmatrix} 1 & 0 & 2 \\ 0 & 1 & 2 \\ 1 & 2 & 0 \end{pmatrix}\begin{pmatrix} 1 & 0 & 2 \\ 0 & 1 & 2 \\ 1 & 2 & 0 \end{pmatrix} - 2\begin{pmatrix} 1 & 0 & 2 \\ 0 & 1 & 2 \\ 1 & 2 & 0 \end{pmatrix} - 5\begin{pmatrix} 1 & 0 & 0 \\ 0 & 1 & 0 \\ 0 & 0 & 1 \end{pmatrix}$$

$$= \begin{pmatrix} 3 & 4 & 2 \\ 2 & 5 & 2 \\ 1 & 2 & 6 \end{pmatrix} - \begin{pmatrix} 2 & 0 & 4 \\ 0 & 2 & 4 \\ 2 & 4 & 0 \end{pmatrix} - \begin{pmatrix} 5 & 0 & 0 \\ 0 & 5 & 0 \\ 0 & 0 & 5 \end{pmatrix}$$

$$= \begin{pmatrix} -4 & 4 & -2 \\ 2 & -2 & -2 \\ -1 & -2 & 1 \end{pmatrix}$$

$$\therefore \quad \text{Inverse } A^{-1} = -\frac{1}{6}\begin{pmatrix} -4 & 4 & -2 \\ 2 & -2 & -2 \\ -1 & -2 & 1 \end{pmatrix} = \begin{pmatrix} 2/3 & -2/3 & 1/3 \\ -1/3 & 1/3 & 1/3 \\ 1/6 & 1/3 & -1/6 \end{pmatrix} \quad \textbf{Ans.}$$

Example 4.12. *Using matrix method, solve the following system of linear equations:*

$$\begin{aligned} 2x - 3y + 5z &= 11 \\ 3x + 2y - 4z &= -5 \\ x + y - 2z &= -3 \end{aligned}$$

Solution: Given equations are

$$\begin{aligned} 2x - 3y + 5z &= 11 \\ 3x + 2y - 4z &= -5 \\ x + y - 2z &= -3 \end{aligned}$$

Writing the equations in the form of a matrix equation

$$\begin{bmatrix} 2 & -3 & 5 \\ 3 & 2 & -4 \\ 1 & 1 & -2 \end{bmatrix}\begin{bmatrix} x \\ y \\ z \end{bmatrix} = \begin{bmatrix} 11 \\ -5 \\ -3 \end{bmatrix} \Rightarrow A \cdot X = B$$

$$|A| = 2(-4+4) + 3(-6+4) + 5(3-2) = 0 - 6 + 5 = -1$$

The cofactors are

$$\begin{array}{lll} A_{11} = 0, & A_{12} = -6 & A_{13} = -1 \\ A_{21} = -1, & A_{22} = -9 & A_{23} = -4 \\ A_{31} = 2, & A_{32} = +23 & A_{33} = -13 \end{array}$$

$\therefore$ The matrix of cofactors is $\begin{bmatrix} 0 & +2 & +1 \\ -1 & -9 & -5 \\ 2 & 23 & 13 \end{bmatrix}$

Adj. $A = \begin{bmatrix} 0 & -1 & 2 \\ 2 & -9 & 23 \\ 1 & -5 & 13 \end{bmatrix} \quad \therefore \quad A^{-1} = \dfrac{1}{-1}\begin{bmatrix} 0 & -1 & 2 \\ 2 & -9 & 23 \\ 1 & -5 & 13 \end{bmatrix}$

$$AX = B \Rightarrow X = A^{-1}B$$

$$\begin{bmatrix} x \\ y \\ z \end{bmatrix} = -1\begin{bmatrix} 0 & -1 & 2 \\ 2 & -9 & 23 \\ 1 & -5 & 13 \end{bmatrix}\begin{bmatrix} 11 \\ -5 \\ -3 \end{bmatrix} = -1\begin{bmatrix} 0+5-6 \\ 22+45-69 \\ 11+25-39 \end{bmatrix} = \begin{bmatrix} 1 \\ 2 \\ 3 \end{bmatrix}$$

$\therefore$ $x = 1, y = 2, z = 3$ **Ans.**

Example 4.13. *Using matrix method, solve the following system of linear equations:*

$$\begin{aligned} x + y + z &= 3 \\ 2x - y + z &= 2 \\ x - 2y + 3z &= 2 \end{aligned}$$

Solution: The system of equations may be written as a matrix equation

$$\begin{bmatrix} 1 & 1 & 1 \\ 2 & -1 & 1 \\ 1 & -2 & 3 \end{bmatrix} \begin{bmatrix} x \\ y \\ z \end{bmatrix} = \begin{bmatrix} 3 \\ 2 \\ 1 \end{bmatrix} \quad \text{or} \quad A \cdot X = B$$

$|A| = 1(-3+2) - 1(6-1) + 1(-4+1) = -9$ being nonsingular, A^{-1} exists

Cofactors of the determinant are

$$\begin{array}{lll} A_{11} = -1, & A_{12} = -5 & A_{13} = -3 \\ A_{21} = -5, & A_{22} = 2 & A_{23} = +3 \\ A_{31} = +2, & A_{32} = +1 & A_{33} = -3 \end{array}$$

$$\text{Matrix of cofactors} = \begin{vmatrix} -1 & -5 & -3 \\ -5 & 2 & 3 \\ +2 & 1 & -3 \end{vmatrix}$$

$$\text{Adjoint} = \begin{vmatrix} -1 & -5 & 2 \\ -5 & 2 & 1 \\ +2 & 3 & -3 \end{vmatrix}$$

$$A^{-1} = \frac{1}{-9} \begin{bmatrix} -1 & -5 & 2 \\ -5 & 2 & 1 \\ -3 & 3 & -3 \end{bmatrix}$$

$$AX = B \Rightarrow X = A^{-1} B = -\frac{1}{9} \begin{bmatrix} -1 & -5 & 2 \\ -5 & 2 & 1 \\ -3 & 3 & -3 \end{bmatrix} \begin{bmatrix} 3 \\ 2 \\ 2 \end{bmatrix}$$

$$\begin{bmatrix} x \\ y \\ z \end{bmatrix} = -\begin{bmatrix} -3 & -10 & +4 \\ -15 & +4 & +2 \\ -9 & +6 & -6 \end{bmatrix} \Rightarrow \frac{1}{9} \begin{bmatrix} x \\ y \\ z \end{bmatrix} = \begin{bmatrix} -9 \\ -9 \\ -9 \end{bmatrix}$$

$\therefore \quad x = 1, y = 1, z = 1$ **Ans.**

Example 4.14. *Using matrix method, solve the following system of linear equations:*

$$\begin{aligned} x + y - z &= 1 \\ 3x + y - 2z &= 3 \\ x - y - z &= -1 \end{aligned}$$

Solution: Writing in matrix notation

$$\begin{bmatrix} 1 & 1 & -1 \\ 3 & 1 & -2 \\ 1 & -1 & -1 \end{bmatrix} \begin{bmatrix} x \\ y \\ z \end{bmatrix} = \begin{bmatrix} 1 \\ 3 \\ -1 \end{bmatrix}$$

or
$$AX = B \Rightarrow X = A^{-1}B$$
$$|A| = 1(-1-2)-(3+2)-1(-3-1)$$
$$= -3+1+4 = 2 \neq 0$$

Hence, matgix A is non-singular and its inverse exists.

Cofacors of the determinant A are:

$A_{11} = (-1-2) = -3$ $\quad A_{12} = -(-3+2) = 1$ $\quad A_{13} = (-3-1) = -4$

$A_{21} = -(-1-1) = 2$ $\quad A_{22} = (-1+1) = 0$ $\quad A_{23} = -(-1-1) = 2$

$A_{31} = (-2+1) = -1$ $\quad A_{32} = -(-3+3) = -1$ $\quad A_{33} = (1-3) = -2$

Hence,
$$A^{-1} = \frac{1}{2}\begin{bmatrix} -3 & 2 & -1 \\ 1 & 0 & -1 \\ -4 & 0 & -2 \end{bmatrix}$$

and
$$A^{-1}B = \frac{1}{2}\begin{bmatrix} -3 & 2 & -1 \\ 1 & 0 & -1 \\ -4 & 2 & -2 \end{bmatrix}\begin{bmatrix} 1 \\ 3 \\ -1 \end{bmatrix} = \frac{1}{2}\begin{bmatrix} -3+6+1 \\ 1+0+1 \\ -4+6+2 \end{bmatrix}$$

$$\begin{bmatrix} x \\ y \\ z \end{bmatrix} = \frac{1}{2}\begin{bmatrix} 4 \\ 2 \\ 4 \end{bmatrix} \quad \text{or} \quad \begin{bmatrix} x \\ y \\ z \end{bmatrix} = \begin{bmatrix} 2 \\ 1 \\ 2 \end{bmatrix}$$

$\therefore$ $\quad x = 2; y = 1; z = 2$ **Ans.**

Example 4.15. *If* $A = \begin{bmatrix} 1 & 2 & 2 \\ 2 & 1 & 2 \\ 2 & 2 & 1 \end{bmatrix}$, *prove that* $A^2 - 4A - 5I = 0$. *Hence, find* A^{-1}.

Solution: $A = \begin{bmatrix} 1 & 2 & 2 \\ 2 & 1 & 2 \\ 2 & 2 & 1 \end{bmatrix}$

$$\therefore \quad A^2 = \begin{bmatrix} 1 & 2 & 2 \\ 2 & 1 & 2 \\ 2 & 2 & 1 \end{bmatrix}\begin{bmatrix} 1 & 2 & 2 \\ 2 & 1 & 2 \\ 2 & 2 & 2 \end{bmatrix}\begin{bmatrix} 1+4+4 & 2+2+4 & 2+4+2 \\ 2+2+4 & 4+1+4 & 4+2+2 \\ 2+4+2 & 4+2+2 & 4+4+1 \end{bmatrix}$$

$$= \begin{bmatrix} 9 & 8 & 8 \\ 8 & 9 & 8 \\ 8 & 8 & 9 \end{bmatrix}$$

$$\therefore \quad A^2 - 4A - 51 = \begin{bmatrix} 9 & 8 & 8 \\ 8 & 9 & 8 \\ 8 & 8 & 9 \end{bmatrix} - \begin{bmatrix} 4 & 8 & 8 \\ 8 & 4 & 8 \\ 8 & 8 & 4 \end{bmatrix} - \begin{bmatrix} 5 & 0 & 0 \\ 0 & 5 & 0 \\ 0 & 0 & 5 \end{bmatrix}$$

$$= \begin{bmatrix} 0 & 0 & 0 \\ 0 & 0 & 0 \\ 0 & 0 & 0 \end{bmatrix} = 0 \quad \text{Proved.}$$

Pre-multiplying $A^2 - 4A - 5I = 0$ by A^{-1} we get

$$A - 4I - 5A^{-1} = 0$$

or
$$A^{-1} = \frac{1}{5}(A - 4I) = \frac{1}{5}\left[\begin{bmatrix} 1 & 2 & 2 \\ 2 & 1 & 2 \\ 2 & 2 & 1 \end{bmatrix} - \begin{bmatrix} 4 & 0 & 0 \\ 0 & 4 & 0 \\ 0 & 0 & 4 \end{bmatrix}\right]$$

or
$$= \frac{1}{5}\begin{bmatrix} -3 & 2 & 2 \\ 2 & -3 & 2 \\ 2 & 2 & 3 \end{bmatrix} \quad \textbf{Ans.}$$

Example 4.16. *Solve the following system of linear equations by means of Cramer's rule.*

$$x + y + z = 7$$
$$x + 2y + 3z = 16$$
$$x + 3y + 4z = 20$$

Solution: Here

$$\Delta = \begin{vmatrix} 1 & 1 & 1 \\ 1 & 2 & 3 \\ 1 & 3 & 4 \end{vmatrix} = \begin{vmatrix} 1 & 0 & 0 \\ 1 & 1 & 2 \\ 1 & 2 & 3 \end{vmatrix}$$

Applying $C_2 \rightarrow (C_2 - C_1)$ and $C_3 \rightarrow (C_3 - C_1)$

$$\Delta = -1 \neq 0 \qquad \ldots(i)$$

$$\Delta x = \begin{vmatrix} 7 & 1 & 1 \\ 16 & 2 & 3 \\ 20 & 3 & 4 \end{vmatrix} = \begin{vmatrix} 0 & 0 & 1 \\ -5 & -1 & 3 \\ -8 & -1 & 4 \end{vmatrix}$$

Applying $C_1 \rightarrow (C_1 - 7C_2)$ and $C_2 \rightarrow (C_2 - C_3)$

$$= 5 - 8 = -3$$

$$\Rightarrow \quad \Delta x = -3 \qquad \ldots(ii)$$

$$\Delta y = \begin{vmatrix} 1 & 7 & 1 \\ 1 & 16 & 3 \\ 1 & 20 & 4 \end{vmatrix} = \begin{vmatrix} 1 & 7 & 1 \\ 0 & 9 & 2 \\ 0 & 13 & 3 \end{vmatrix}$$

Applying $R_1 \to R_1 - R_1$ and $R_3 \to R_3 - R_1$

$$= 27 - 26 = 1$$

$$\Rightarrow \quad \Delta y = 1 \qquad ...(iii)$$

$$\Delta z = \begin{vmatrix} 1 & 1 & 7 \\ 1 & 2 & 16 \\ 1 & 3 & 20 \end{vmatrix} = \begin{vmatrix} 1 & 1 & 7 \\ 0 & 1 & 9 \\ 0 & 2 & 13 \end{vmatrix}$$

Applying $R_2 \to R_2 - R_1$ and $R_3 \to R_3 - R_1$

$$= 13 - 18 = -5$$

$$\Rightarrow \quad \Delta z = -5 \qquad ...(iv)$$

$\therefore$ In view of eqns. (*i*), (*ii*), (*iii*) and (*iv*)

$$x = \frac{\Delta x}{\Delta} = \frac{-3}{-1} = 3,\ y = \frac{\Delta y}{\Delta} = \frac{1}{-1} = -1,\ z = \frac{\Delta z}{\Delta} = \frac{-5}{-1} = 5$$

Hence,

$$x = 3, y = -1, z = 5 \quad \textbf{Ans.}$$

Example 4.17. *Solve the following system of linear equations by Cramer's Rule.*

$$3x - 4y - z = 2$$
$$6x + 6y + 3z = 7$$
$$9x - 8y - 5z = 0$$

Solution:

$$\text{Determinant, } \Delta = \begin{vmatrix} 3 & -4 & -1 \\ 6 & 6 & 3 \\ 9 & -8 & -5 \end{vmatrix}$$

Applying $C_1 \to (C_1 + 3C_3)$ and $C_2 \to (C_2 - 4C_3)$

$$\begin{vmatrix} 0 & 0 & -1 \\ 15 & -6 & 3 \\ -6 & 12 & -3 \end{vmatrix}$$

$$= -(180 - 36) = -144$$

$$\Rightarrow \quad \Delta = -144 \neq 0 \qquad ...(i)$$

Similarly,

$$\Delta x = \begin{vmatrix} 2 & -4 & -1 \\ 7 & 6 & 3 \\ 0 & -8 & -5 \end{vmatrix} = -96$$

$$\Rightarrow \quad \Delta x = -96 \qquad ...(ii)$$

$$\Delta y = \begin{vmatrix} 3 & 2 & -1 \\ 6 & 7 & 3 \\ 9 & 0 & -5 \end{vmatrix} = 72$$

$$\Rightarrow \quad \Delta x = 72 \quad \quad ...(iii)$$

$$\Delta z = \begin{vmatrix} 3 & -4 & 2 \\ 6 & 6 & 7 \\ 9 & -8 & 0 \end{vmatrix} = -288$$

$$\Rightarrow \quad \Delta z = -288 \quad \quad ...(iv)$$

$\therefore$ In view of eqns. (*i*), (*ii*), (*iii*) and (*iv*) we get

$$x = \frac{\Delta x}{\Delta} = \frac{-96}{-144} = \frac{2}{3}; \quad y = \frac{\Delta y}{\Delta} = \frac{72}{-144} = \frac{-1}{2}$$

$$z = \frac{\Delta z}{\Delta} = \frac{-288}{-144} = 2$$

Hence

$$x = \frac{2}{3}, \quad y = \frac{-1}{2}, \quad z = 2 \quad \textbf{Ans.}$$

Example 4.18. *If* $\begin{bmatrix} 1 & 1 & 1 \\ 1 & 2 & -3 \\ 2 & -1 & 3 \end{bmatrix}$, *find* A^{-1}.

$$x + y + 2z = 0$$
$$x + 2y - z = 9$$
$$x - 3y - 3z = -14$$

Solution: $$|A| = \begin{vmatrix} 1 & 1 & 1 \\ 1 & 2 & -3 \\ 2 & -1 & 3 \end{vmatrix} = 1(6-3) - 1(3+6) + 1(-1-4)$$

$$= 3 - 9 - 5 = -11 \neq 0 \quad \text{Hence, } A^{-1} \text{ exists.}$$

$A_{11} = 6 - 3 = 3$ $\quad A_{12} = -(3 + 6) = -9$ $\quad A_{13} = 1(-1-4) = -5$

$A_{21} = -1(3 + 1) = -4$ $\quad A_{22} = (3 - 2) = 1$ $\quad A_{23} = -(-1-2) = 3$

$A_{31} = (-3 - 2) = -5$ $\quad A_{32} = -(-3-1) = 4$ $\quad A_{33} = (2 - 1) = 1$

$$\therefore \text{Matrix of cofactors} = \begin{bmatrix} 3 & -9 & -5 \\ -4 & 1 & 3 \\ -5 & 4 & 1 \end{bmatrix}$$

Adjoint $A = \begin{bmatrix} 3 & -4 & 5 \\ -9 & 1 & 4 \\ -5 & 3 & 1 \end{bmatrix}$

$$\therefore \quad A^{-1} = \frac{1}{(-11)}\begin{bmatrix} 3 & -4 & 5 \\ -9 & 1 & 4 \\ -5 & 3 & 1 \end{bmatrix} = \frac{1}{11}\begin{bmatrix} -3 & 4 & 5 \\ 9 & -1 & -4 \\ 5 & -3 & -1 \end{bmatrix}$$

The given equations can be written as

$$A^{-1}X = B \Rightarrow X = (A')^{-1}B = (A^{-1})'B$$

$$\begin{bmatrix} x \\ y \\ z \end{bmatrix} = \frac{1}{11}\begin{bmatrix} -3 & 9 & 5 \\ 4 & -1 & -3 \\ 5 & -4 & -1 \end{bmatrix}\begin{bmatrix} 0 \\ 9 \\ -11 \end{bmatrix} = \frac{1}{11}\begin{bmatrix} 0 & 8 & -70 \\ 0 & -9 & +42 \\ 0 & -36 & +14 \end{bmatrix} = \frac{1}{11}\begin{bmatrix} 11 \\ 33 \\ -22 \end{bmatrix}$$

$$\therefore \quad \begin{bmatrix} x \\ y \\ z \end{bmatrix} = \begin{bmatrix} 1 \\ 3 \\ -2 \end{bmatrix} \Rightarrow x = 1, y = 3, z = -2 \quad \textbf{Ans.}$$

Example 4.19. *Using matrices solve the following system of equations:*

$$x + \frac{2}{y} + 3xz = 1$$

$$2x - \frac{4}{y} - 3xz = 3$$

$$3x + \frac{6}{y} + 2xz = 4$$

Solution: Let $\frac{1}{y} = u; xz = V$

The given equations reduce to:

$$x + 2u + 3V = -1$$

$$2x - 4u - 3V = 3$$

$$6x + 6u - 2V = 4$$

Applying the matrix formula, we get

$$\begin{bmatrix} 1 & 2 & 3 \\ 2 & -4 & -3 \\ 3 & 6 & 2 \end{bmatrix}\begin{bmatrix} x \\ u \\ V \end{bmatrix} = \begin{bmatrix} -1 \\ 3 \\ 4 \end{bmatrix} \text{ i.e. } AX = B$$

$$|A| = \begin{bmatrix} 1 & 2 & 3 \\ 2 & -4 & -3 \\ 3 & 6 & 2 \end{bmatrix}$$

$$= 1(-8+18) - 2(4+9) + 3(12+12)$$

$$= 10 - 26 + 72 = 56$$

$$\text{Adjoint } A = \begin{bmatrix} 10 & 14 & 6 \\ -13 & -7 & 9 \\ 24 & 0 & -8 \end{bmatrix}$$

$$A^{-1} = \frac{1}{56}\begin{bmatrix} 10 & 14 & 6 \\ -13 & -7 & 9 \\ 24 & 0 & -8 \end{bmatrix}$$

$$X = A^{-1}B = \frac{1}{56}\begin{bmatrix} 10 & 14 & 6 \\ -13 & -7 & 9 \\ 24 & 0 & -8 \end{bmatrix}\begin{bmatrix} -1 \\ 3 \\ 4 \end{bmatrix} = \begin{bmatrix} 1 \\ 1/2 \\ -1 \end{bmatrix}$$

$$x = 1;\ u = \frac{1}{2};\ V = -1$$

$$u = \frac{1}{2} \Rightarrow \frac{1}{y} = \frac{1}{2} \Rightarrow y = 2$$

$$V = -1 \Rightarrow xz = -1 \Rightarrow z = -1$$ **Ans.**

Example 4.20. *Find the eigen values and eigenvectors of the matrix*

$$\begin{bmatrix} 5 & 4 \\ 1 & 2 \end{bmatrix}$$

Solution: Characteristic equation of the given matrix:

$$\begin{bmatrix} 5-\lambda & 4 \\ 1 & 2-\lambda \end{bmatrix} = 0$$

$$(5-\lambda)(2-\lambda) - 4 = 0$$

$$10 - 5\lambda - 2\lambda + \lambda^2 - 4 = 0$$

$$\lambda^2 - 7\lambda + 6 = 0$$

∴ Value of $\lambda = 6, 1$

when $\lambda = 6$

$$\begin{bmatrix} -1 & 4 \\ 1 & -4 \end{bmatrix}\begin{bmatrix} x_1 \\ x_2 \end{bmatrix} = \begin{bmatrix} 0 \\ 0 \end{bmatrix}$$

$$-x_1 + 4x_2 = 0 \quad x_1 = 4x_2 \quad x_2 = 1 \quad x_1 = 4$$
$$x_1 - 4x_2 = 0 \quad x_1 = 4x_2 \quad x_1 = 1 \quad x_2 = 4$$

when $\lambda = 1$

$$\begin{bmatrix} 4 & 4 \\ 1 & 1 \end{bmatrix}\begin{bmatrix} x_1 \\ x_2 \end{bmatrix} = \begin{bmatrix} 0 \\ 0 \end{bmatrix}$$

$$x_1 = x_2 \quad x_1 = 1 \quad x_2 = 1$$

Eigen vectors are [1, 4] [1, 1]

Example 4.21. *Find the eigen values and eigen vectors of*

$$A = \begin{pmatrix} 6 & -2 & 2 \\ -2 & 3 & -1 \\ 2 & -1 & 3 \end{pmatrix}$$

Also find the eigenvalues of A_t^S ·

Solution: $\begin{vmatrix} 6-\lambda & -2 & 2 \\ -2 & 3-\lambda & -1 \\ 2 & -1 & 3-\lambda \end{vmatrix} = 0$

$[(6 - \lambda)\ [(3 - \lambda)\ (3 - \lambda) - 1] - 2[-2(3 - \lambda) - (-2)] + 2[2 \ (3 - \lambda)2 = 0$

or $\quad -\lambda^3 + 12\lambda^2 - 36\lambda + 32 = 0$

Let $\lambda = -2$ be one of the roots.

$$\begin{array}{r} -\lambda^2 + 10\lambda - 16 \\ \lambda + 2) \overline{-\lambda^3 + 12\lambda^2 - 36\lambda + 32} \\ -\lambda^3 - 2\lambda^2 \qquad\qquad \\ -10\lambda^2 - 36\lambda \qquad \\ 10\lambda^2 - 20\lambda \qquad \\ -16\lambda + 32 \\ \underline{-16\lambda - 32} \\ 00 \quad 00 \end{array}$$

$$\lambda = -2 \text{ is one of the roots.}$$

Other roots are

$$\lambda = -8 \text{ and } \lambda = 2$$

∴ Eigen values are 2, – 2, – 8

(*i*) Eigen vector corresponding to 2 is

$$\begin{bmatrix} 4 & -2 & 2 \\ -2 & 1 & -1 \\ 2 & -1 & 1 \end{bmatrix} \begin{bmatrix} x_1 \\ x_2 \\ x_3 \end{bmatrix} = 0$$

$$4x_1 - 2x_2 + 2x_3 = 0$$
$$-2x_1 - x_2 - x_3 = 0$$
$$2x_1 - x_2 + x_3 = 0$$

One possible solution is

$$x_1 = 1 \quad x_2 = 4 \quad x_3 = 2$$

$$\therefore \quad \begin{bmatrix} 1 \\ 4 \\ 2 \end{bmatrix} \text{ is an eigen vector}$$

(*ii*) Eigen vector corresponding to – 2 is

$$\begin{bmatrix} 6-(-2) & -2 & 2 \\ -2 & 3-(-2) & -1 \\ 2 & -1 & 3-(-2) \end{bmatrix} \begin{bmatrix} x_1 \\ x_2 \\ x_3 \end{bmatrix} = 0$$

$$\begin{bmatrix} 8 & -2 & 2 \\ -2 & 5 & -1 \\ 2 & -1 & 5 \end{bmatrix} \begin{bmatrix} x_1 \\ x_2 \\ x_3 \end{bmatrix} = 0$$

$$8x_1 - 2x_2 + 2x_3 = 0$$
$$-2x_1 + 5x_2 - x_3 = 0$$
$$2x_1 - x_2 + 5x_3 = 0$$

From first two equations

$$\frac{x_1}{2-10} = \frac{-x_2}{-8+4} = \frac{x_3}{40-4}$$

$$\frac{x_1}{-8} = \frac{-x_2}{-4} = \frac{x_3}{36}$$

$$\begin{bmatrix} -8 \\ -4 \\ 36 \end{bmatrix} \text{ is another eigen vector}$$

(*iii*) Eigen vector corresponding to – 8 is

$$\begin{bmatrix} 5-(-8) & -2 & 2 \\ -2 & 3-(-8) & -1 \\ 2 & -1 & 3-(-8) \end{bmatrix} \begin{bmatrix} x_1 \\ x_2 \\ x_3 \end{bmatrix} = 0$$

$$\begin{bmatrix} 14 & -2 & 2 \\ -2 & 11 & -1 \\ 2 & -1 & 11 \end{bmatrix} \begin{bmatrix} x_1 \\ x_2 \\ x_3 \end{bmatrix} = 0$$

$$14x_1 - 2x_2 + 2x_3 = 0$$

$$-2x_1 - 11x_2 - x_3 = 0$$

$$2x_1 - x_2 + 11x_3 = 0$$

From first two equations

$$\frac{x_1}{2-22} = \frac{-x_2}{-14+4} = \frac{x_3}{154-4}$$

$$\frac{x_1}{-20} = \frac{-x_2}{-10} = \frac{x_3}{150}$$

$\therefore \quad \begin{bmatrix} -2 \\ 1 \\ 15 \end{bmatrix}$ is another eigen vector

Eigen values for A are 2, – 2 and – 8

$\therefore$ Eigen values of A^2 are 4, – 4, – 16

A^3 are 8, – 8, – 32

$\vdots$

A^8 are 256, – 256, – 1024

Example 4.22. *Determine the eigen values and eigen vectors for the following matrix:*

$$A = \begin{bmatrix} 0 & 1 & 1 \\ 1 & 0 & 1 \\ 1 & 1 & 0 \end{bmatrix}$$

Solution: Given matrix.

$$A = \begin{bmatrix} 0 & 1 & 1 \\ 1 & 0 & 1 \\ 1 & 1 & 0 \end{bmatrix}$$

Now, $$I = \begin{bmatrix} 1 & 0 & 0 \\ 0 & 1 & 0 \\ 0 & 0 & 1 \end{bmatrix} \text{ and } \lambda I = \begin{bmatrix} \lambda & 0 & 0 \\ 0 & \lambda & 0 \\ 0 & 0 & \lambda \end{bmatrix}$$

$$\therefore \quad A - \lambda I = \begin{bmatrix} 0-\lambda & 1 & 1 \\ 1 & 0-\lambda & 1 \\ 1 & 1 & 0-\lambda \end{bmatrix} = 0$$

The characteristic equation is obtained by expanding the matrix

$$-\lambda(\lambda - 1) - 1(-\lambda - 1) + (1 + \lambda) = 0$$
$$-\lambda^3 + \lambda + \lambda + 1 + 1 + \lambda = 0$$
$$-\lambda^3 + 3\lambda + 2 = 0$$

The roots of the characteristic equation matrix are the eigen values.

By putting $\lambda = -1$ we find $+ 1 - 3 + 2 = 0$ *i.e.*, $(\lambda + 1)$ is a factor

$$(\lambda + 1)(-\lambda^2 + \lambda + 2) = 0$$
$$(\lambda + 1)(\lambda + 1)(-\lambda + 2) = 0$$

or $$\lambda = -1, -1, 2.$$

Hence, the eigen values are $-1, -1$, and 2 **Ans.**

Corresponding eigen vectors are calculated as under:

For $\lambda = -1$,

$$\begin{bmatrix} 0+1 & 1 & 1 \\ 1 & 0+1 & 1 \\ 1 & 1 & 0+1 \end{bmatrix} \begin{bmatrix} x_1 \\ x_2 \\ x_3 \end{bmatrix} = 0 \Rightarrow \begin{bmatrix} 1 & 1 & 1 \\ 1 & 1 & 1 \\ 1 & 1 & 1 \end{bmatrix} \begin{bmatrix} x_1 \\ x_2 \\ x_3 \end{bmatrix} = 0$$

or $$x_1 + x_2 + x_3 = 0$$
$$x_1 + x_2 + x_3 = 0$$
$$x_1 + x_2 + x_3 = 0$$

Apparently, Eigen vectors are: [0 , 0, 0] **Ans.**

For $\lambda = 2$ we get

$$\begin{bmatrix} 0-1 & 1 & 1 \\ 1 & 02 & 1 \\ 1 & 1 & 02 \end{bmatrix} \begin{bmatrix} x_1 \\ x_2 \\ x_3 \end{bmatrix} = 0 \Rightarrow \begin{bmatrix} -2 & 1 & 1 \\ 1 & -2 & 1 \\ 1 & 1 & -2 \end{bmatrix} \begin{bmatrix} x_1 \\ x_2 \\ x_3 \end{bmatrix} = 0$$

or $$-2x_1 + x_2 + x_3 = 0 \qquad \ldots(i)$$
$$x_1 + -2x_2 + x_3 = 0 \qquad \ldots(ii)$$
$$x_1 + x_2 - 2x_3 = 0 \qquad \ldots(iii)$$

For eqns (*i*) and (*ii*) we get

$$-2x_1 + x_2 = x_1 - 2x_2$$
$$x_1 = x_2$$

For eqns (*ii*) and (*iii*) we get

$$-2x_2 + x_3 = x_2 - 2x_3$$
$$x_2 = x_3$$
$$\Rightarrow \quad x_1 = x_2 = x_3$$

The eigen vectors are [0, 0, 0] **Ans.**

Example 4.23. *Find the eignvalues and their corresponding eign vectors for the matrix*

$$A = \begin{bmatrix} 5 & 2 & 1 \\ 2 & 5 & 1 \\ 2 & 1 & 5 \end{bmatrix}$$

Solution: Given matrix

$$A = \begin{bmatrix} 4 & 2 & 1 \\ 2 & 5 & 1 \\ 2 & 1 & 5 \end{bmatrix}$$

Let λ be any scalar quantity. The characteristic matrix is given by

$$\begin{bmatrix} 5-\lambda & 2 & 1 \\ 2 & 5-\lambda & 1 \\ 2 & 1 & 5-\lambda \end{bmatrix}$$

The characteristic equation is obtained by expanding the characteristic matrix

$$(5-\lambda)[25 - 10\lambda + \lambda^2 - 1] - 2[2(5-\lambda) - 2] + 1[2 - 2(5-\lambda)] = 0$$
$$(5-\lambda)(24 - 10\lambda + \lambda^2) - 2(10 - 2\lambda - 2) + 1(2 - 10 + 2\lambda) = 0$$
$$120 - 50\lambda + 5\lambda^2 - 24\lambda + 10\lambda^2 - \lambda^3 - 2(8 - 2\lambda) + 1(-8 + 2\lambda) = 0$$
$$120 - 50\lambda + 6\lambda^2 - 24\lambda + 10\lambda^2 - \lambda^3 - 16 + 4\lambda - 8 + 2\lambda = 0$$
$$-\lambda^3 + 15\lambda^2 - 68\lambda + 96 = 0$$

By putting λ = 3 we get

$$-27 + 135 - 204 + 96 = 0 \quad \text{i.e., } (\lambda - 3) \text{ is one of the factors.}$$

is $$(\lambda - 3)(-\lambda^2 + 2\lambda - 32) = 0$$

or $$(\lambda - 3)(\lambda - 4)(\lambda - 8) = 0$$

Hence, the eign values are: 3, 4, 8. **Ans.**

Calculation of the corresponding eign vectors of the matrix.

For λ = 3, (1)

$$\begin{bmatrix} 5-3 & 2 & 1 \\ 2 & 5-3 & 1 \\ 2 & 1 & 5-3 \end{bmatrix} \begin{bmatrix} x_1 \\ x_2 \\ x_3 \end{bmatrix} = 0 \Rightarrow \begin{bmatrix} 2 & 2 & 1 \\ 2 & 2 & 1 \\ 2 & 1 & 2 \end{bmatrix} \begin{bmatrix} x_1 \\ x_2 \\ x_3 \end{bmatrix} = 0$$

$$2x_1 + 2x_2 + x_3 = 0 \quad ...(1)$$
$$2x_1 + 2x_2 + x_3 = 0 \quad ...(2)$$
$$2x_1 + x_2 + 2x_3 = 0 \quad ...(3)$$

From eqns (1) and (2) we get

$$2x_2 + x_3 = 2x_2 + x_3$$

or
$$x_2 = -x_3$$

We know that for an eign vector, $x_1^2 + x_2^2 + x_3^2 = 1$

By choosing $x_1 = \frac{1}{\sqrt{2}}$ and substituting weget

$$\frac{1}{2} + \frac{1}{2} + 0 = 1$$

Hence, for $\lambda = 3$, vector is $\left[0, \frac{1}{\sqrt{2}}, -\frac{1}{\sqrt{2}}\right]$

For $\lambda = 4$, the eigen vector is obtained as under:

$$\begin{bmatrix} 5-4 & 2 & 1 \\ 2 & 5-4 & 1 \\ 2 & 1 & 5-4 \end{bmatrix} \begin{bmatrix} x_1 \\ x_2 \\ x_3 \end{bmatrix} = 0$$

$$x_1 + 2x_2 + x_3 = 0 \quad ...(i)$$
$$2x_1 + x_2 + x_3 = 0 \quad ...(ii)$$
$$2x_1 + x_2 + x_3 = 0 \quad ...(iii)$$

From eqn (*i*) and (*ii*) we get

$$x_1 + 2x_2 = 2x_1 + x_2$$

or
$$x_1 = x_2$$

By choosing $x_1 = x_2 = \frac{1}{\sqrt{2}}$ and substituting in eqn. $x_1^2 + x_2^2 + x_3^2 = 1$,

$$\frac{1}{2} + \frac{1}{2} + 0 = 1$$

i.e., the eigen vector is $\left[\frac{1}{\sqrt{2}}, \frac{1}{\sqrt{2}}, 0\right]$ **Ans.**

For $\lambda = 8$, the vector is obtained as under:

$$\begin{bmatrix} 5-8 & 2 & 1 \\ 2 & 5-8 & 1 \\ 2 & 1 & 5-4 \end{bmatrix} \begin{bmatrix} x_1 \\ x_2 \\ x_3 \end{bmatrix} = 0$$

$$-3x_1 + 2x_2 + x_3 = 0 \quad ...(a)$$

$$2x_1 - 3x_2 + x_3 = 0 \quad ...(b)$$
$$2x_1 + x_2 - 3x_3 = 0 \quad ...(c)$$

From eqns (*a*) and (*b*) we get

$$-5x_1 + 5x_2 = 0$$

$$\therefore \quad x_1 = x_2$$

Now choosing $x_1 = \frac{1}{\sqrt{2}}, x_2 = \frac{1}{\sqrt{2}}$ or $\frac{1}{2} + \frac{1}{2} + 0 = 1$

The third eigen vector is $\left[\frac{1}{\sqrt{2}}, \frac{1}{\sqrt{2}}, 0\right]$ **Ans.**

Example 4.24. *Find the Eigen values and the eigen vectors for the matrix*

$$A = \begin{bmatrix} 5 & 0 & -1 \\ 0 & -2 & 0 \\ -1 & 0 & 5 \end{bmatrix}$$

Solution: Its characteristic equation is

$$[A - \lambda I] = 0$$

i.e.,
$$= \begin{bmatrix} 5-\lambda & 0 & -1 \\ 0 & -2-\lambda & 0 \\ -1 & 0 & 5-\lambda \end{bmatrix} = 0$$

On expansion of the determinant, we get

$$(5 - \lambda)\,[-(2 + \lambda)\,(5 - \lambda)] - 1\,(0 - (2 + \lambda)) = 0$$
$$-(5 - \lambda)^2\,(\lambda + 2) + (\lambda + 2) = 0$$
$$-(\lambda + 2)\,(25 - 10\lambda + \lambda^2) + (\lambda + 2) = 0$$
$$-(\lambda + 2)\,(25 - 10\lambda + \lambda^2 - 1) = 0$$
$$(\lambda + 2)\,(\lambda^2 - 10\lambda + 24) = 0$$
$$(\lambda + 2)\,(x - 6)\,(x - 4) = 0$$

$\therefore$ The roots are: $\lambda = -2, 6, 4$

and Eigen values are:

$$\left.\begin{array}{l} \lambda_1 = -2 \\ \lambda_2 = 6 \\ \lambda_3 = 4 \end{array}\right\} \textbf{Ans.}$$

Eigen vectors

(*i*) Eign vector for $\lambda = 2$

$$\begin{bmatrix} 5-2 & 0 & -1 \\ 0 & -2+2 & 0 \\ -1 & 0 & 5+2 \end{bmatrix} \begin{bmatrix} x_1 \\ x_2 \\ x_3 \end{bmatrix} = \begin{bmatrix} 0 \\ 0 \\ 0 \end{bmatrix}$$

$$\begin{bmatrix} 7 & 0 & -1 \\ 0 & 0 & 0 \\ -1 & 0 & 7 \end{bmatrix} \begin{bmatrix} x_1 \\ x_2 \\ x_3 \end{bmatrix} = \begin{bmatrix} 0 \\ 0 \\ 0 \end{bmatrix}$$

or
$$7x_1 - x_3 = 0$$
$$-x_1 + 7x_3 = 0$$

Example 4.25. *Find the eign values and their corresponding eign vectors for the matrix:*

$$A = \begin{bmatrix} 5 & 2 & 1 \\ 2 & 5 & 1 \\ 2 & 1 & 5 \end{bmatrix}$$

Solution:

$$A = \begin{bmatrix} 5 & 2 & 1 \\ 2 & 5 & 1 \\ 2 & 1 & 5 \end{bmatrix}$$

Let λ be any scalar quantity. The characteristic matrix is as under

$$\begin{bmatrix} 5-\lambda & 2 & 1 \\ 2 & 5-\lambda & 1 \\ 2 & 1 & 5-\lambda \end{bmatrix}$$

The characteristic equation is

$$-\lambda^3 + 15\lambda^2 - 68\lambda + 96 = 0$$

or $(\lambda - 3)(-\lambda^3 + 12\lambda - 32)$

$(\lambda - 3)(-\lambda^2 + 8\lambda + 4\lambda - 32)$

$(\lambda - 3)(\lambda(-\lambda + 8) - 4(-\lambda + 8)$

$(\lambda - 3)(-\lambda + 8)(\lambda - 4) = 0$

i.e., $\lambda = 3$, $\lambda = 4$, $\lambda = 8$ are roots.

Hence eigen values are 3, 4, 8 **Ans.**

For $\lambda_1 = 3$, the eign vector is obtained as under:

$$\begin{bmatrix} 5-3 & 2 & 1 \\ 2 & 5-3 & 1 \\ 2 & 1 & 5-3 \end{bmatrix} \begin{bmatrix} x_1 \\ x_2 \\ x_3 \end{bmatrix} = 0 \Rightarrow \begin{bmatrix} 2 & 2 & 1 \\ 2 & 2 & 1 \\ 2 & 1 & 2 \end{bmatrix} \begin{bmatrix} x_1 \\ x_2 \\ x_3 \end{bmatrix} = 0$$

$$2x_1 + 2x_2 + x_3 = 0 \quad ...(1)$$
$$2x_1 + 2x_2 + x_3 = 0 \quad ...(2)$$
$$2x_1 + x_2 + 2x_3 = 0 \quad ...(3)$$

From eqns (1) and (2) we get

$$2x_2 + x_3 = 2x_2 + x_3$$

or $$x_2 + x_3 = x_2 + x_3$$

or $$x_2 = -x_3$$

We know that in an eign vector $x_1^2 + x_2^2 + x_3^2 = 1$

Pay charsing $x_1 = \frac{1}{\sqrt{2}}$ we get

eign vector as $\left[0, \frac{1}{\sqrt{2}}, -\frac{1}{\sqrt{2}}\right]$

For $\lambda_2 = 4$, the eign vector is obtained as under

$$x_1 + 2x_2 + x_3 = 0 \quad ...(i)$$

$$2x_1 + x_2 + x_3 = 0 \quad ...(ii)$$

$$2x_1 + x_2 + x_3 = 0 \quad ...(iii)$$

From eqns (*i*) and (*ii*) we get

$$4x_2 + 2x_3 = x_2 + x_3$$

$$3x_3 = x_3$$

$$\frac{x_2}{1} = \frac{x_3}{3}$$

Again; $x_1 + 2x_2 = 2x_1 + x_2$

$$-x_1 = -x_2 \text{ or } x_1 = x_2$$

By choosing $x_1 = x_2 = \frac{1}{\sqrt{2}}$ we get

$$\left(\frac{1}{\sqrt{2}}\right)^2 + \left(\frac{1}{\sqrt{2}}\right)^2 + x = 1$$

$$\frac{1}{2} + \frac{1}{2} + x = 1$$

i.e., $x = 0$

The next given vectors is

$$\left[\frac{1}{\sqrt{2}}, \frac{1}{\sqrt{2}}, 0\right] \quad \textbf{Ans.}$$

Example 4.26. *Find the eigen values and eigen vectors of the matrix*

$$A = \begin{bmatrix} 6 & -2 & 2 \\ -2 & 3 & -1 \\ 2 & -1 & 3 \end{bmatrix}$$

Solution: The characteristic equation is $(A - \lambda I) = 0$

$$\begin{bmatrix} 6-\lambda & -2 & 2 \\ -2 & 3-\lambda & -1 \\ 2 & -1 & 3-\lambda \end{bmatrix} = 0$$

$(6 - \lambda)[(3 - \lambda)^2 - 1] + 2[(-6 + 2\lambda) + 2] \times 2[2 - (6 - 2\lambda)]$

$(6 - \lambda)(9 - 6\lambda + \lambda^2 - 1) + 2(-6 + 2\lambda + 2) + 2(2 - 6 + 2\lambda)$

$(6 - \lambda)(8 - 6\lambda + \lambda^2) - 8 + 4\lambda - 8 + 4\lambda$

$48 - 36\lambda + 6\lambda^2 - \cancel{8\lambda} + 6\lambda^2 - \lambda^3 - 16 + \cancel{8\lambda}$

$$-\lambda^3 + 12\lambda^2 - 36\lambda + 32 = 0 \qquad \text{...}(i)$$

or $$\lambda^3 - 12\lambda^2 + 36\lambda - 32 = 0$$

Putting $\lambda = 2$, we get

$$-8 + 48 - 72 + 32 = 0$$

i.e. $(\lambda - 2)$ is the factor of the equation (i)

$$\lambda - 2)\cancel{-\lambda^3} + 12\lambda^2 - 36\lambda + 32(-\lambda^2 + 10\lambda + 16$$

$$\cancel{-\lambda^3} + 2\lambda^2$$

$$+ \quad -$$

$$+10\lambda^2 - 36\lambda$$

$$+10\lambda^2 - 20\lambda$$

$$- \quad +$$

$$-16\lambda - \cancel{32}$$

$$-16\lambda - \cancel{32}$$

$$+$$

$$0$$

Now $$-\lambda^2 + 10\lambda + 16 = \lambda^2 + 8\lambda + 2\lambda + 16 = 0$$

$$\lambda(\lambda - 8) - 2(\lambda - 8) = 0$$

or $$\lambda = 8, \lambda = 2$$

Hence roots are $\lambda = 2, 2, 8$

❑❑❑

5 Numerical Analysis

5.1 INTRODUCTION

In applied mathematics all kinds of numbers and formulae are used. The normal methods of numerical calculations, are generally inadequate for handling the numerical side of applied mathematics. This is due to the fact that the numerical data used in solving the problems of everyday life, are usually not exact and also, the numbers expressing such data are not exact. Such a data may be true to two, three or more figures. We may categorically state that not only the data of practical problems are approximate, the processes employed to find the desired result are also approximate.

The approximate calculation may be defined as the one which involves approximate data, approximate methods or both. The error in a computed result may therefore be either due to errors in the data or errors due to calculations or both. The errors in the data cannot be remedied but those in the method of calculation, can be made as small as one desires. To achieve this, the calculations are carried out in such a manner that the errors of calculations are negligible. All the numerical calculations are approximate and it is for the computer to adopt a suitable method so that the results consistent with the data, with a minimum labour, are obtained. We shall now discuss the methods relating to approximate calculations and the methods for estimating the accuracy of the results obtained.

Approximation. From the discussion of approximate computation, it is convenient to make a distinction between numbers which are absolutely exact and those which express approximate values.

Let us now consider the following two series of numbers:

(*a*) 1, 2, 3, 4, 5, 6, 7, 8, 9, 0

(*b*) $\sqrt{2}, \sqrt{3}$, π, e, etc.

In (*a*) series, the numbers represent exact numbers as there is no approximation or uncertainty whatsoever.

In (*b*) series, the numbers are exact but these cannot be expressed by finite number of digits. In digital forms, these numbers are:

$$\sqrt{2} = 1.4142136$$

$$\sqrt{3} = 1.7320508$$

$$\pi = 3.1415927$$

$$e = 2.7183$$

These numbers being only approximation to the true numbers, are approximate numbers.

An approximate number may be defined as a number which is used as an approximation to an exact number and differs slightly from its exact value.

Significant figures. In decimal system, the digits 1, 2, 3, 4, 5, 6, 7, 8, 9 are known as **significant figures.** The digit 0 (zero) is also a significant figure if it is used either to fix a decimal point or to fill the places of discarded digits in any number.

Study the following numbers and their significant digits:

Number	Significant digits
(*a*) 0.00125	1, 2, 5
(*b*) 0.01257	1, 2, 5, 7
(*c*) 0.12068	1,2,0,6,8
(*d*) 1.20439	1, 2, 0, 4, 3, 9

In numbers (*a*) and (*b*), the zeros are used only to fix the decimal point. These zeroes are not significant digits as they do not appear in between other significant digits of the numbers.

In some numbers, zeros may appear on the right side of some other significant digits as in the following numbers:

24500, 57320, 836500, etc.

In such cases, there is no indication to show that the zeros are significant or not. Such an ambiguity can be removed by writing the numbers in the powers of 10 notation.

The number 24500 can be written as

2.45×10^4, 2.450×10^4, 2.4500×10^4

Now, total number of significant figures may be indicated by the factor appearing on the left.

5.2 ROUNDING OF NUMBERS

Consider the division of 5 by 3, *i.e.*,

$$5/3 = 1.66666...$$

Here, the quotient never terminates to an exact value. To use such a number in practical computation, we can get the quotient to a manageable form as per the expected accuracy. The quotient of 5/3 may be accepted as

1.6, 1.66, 1.666, 1.6666, ...

The process of cutting off superfluous digits and retaining only the desired digits, is called **'rounding off'.** This is done so as to cause the least possible error in the computation.

For an example, the values of π after rounding off to three, four, five and six figures are:

3.14, 3.142, 3.1416, 3.14159.

Rules of rounding off numbers. The rules which are followed for rounding off numbers are as under:

(*i*) Discard all the digits to the right of the nth place if n is the desired number of significant figures.

(*ii*) If the discarded number is less than half a unit in the nth place, accept the nth digit as it is.

(*iii*) If the discarded number is greater than half a unit in the nth place, add 1 to the nth digit.

(*iv*) If the discarded number is exactly half a unit in the nth place, accept the nth digit as it is, provided it is even number and add 1 if it is an odd number. It is cutomary to round off the nth digit to an even number in such cases.

By applying the above rules, the rounding off is said to be correct to n significant figures. The errors due to rounding off are largely cancelled by one another.

Truncation Error. The difference between a given number and the number obtained after truncation of a specified number of digits from the given number without rounding off, the last retained digit, is called *truncation error.*

In numerical analysis, the results obtained, contain a large number of digits and the requirement is usually upto a few digits. These numerical results have to be reduced to the required number of digits either by rounding off the numbers or by truncating these numbers. The error caused by truncating the numbers, is called a *truncation error*.

If a number x is truncated to x_1 the truncation error

$$E = x - x_1$$

Reason for rounding to an even. In case the discarded number is exactly half a unit in the nth place, the digit is made an even because of achieving a better accuracy in computation. An even digit can be completely divided by 2 and other even numbers as well as by odd numbers. On the other hand, an odd number is never divisible by 2 and other even numbers and is divisible by only few odd numbers. Thus, the rule that the last digit be an even digit, increases the accuracy in computation.

Example 5.1. *Round off the following numbers correct to four significant figures:*

(*i*) *23.5732* (*ii*) *4.39476* (*iii*) *3.85231*

(*iv*) *15.6483* (*v*) *75384* (*vi*) *61.553.*

Solution: The rounded off numbers are shown against each.

Number	Rounded number
(*i*) 23.5732	23.57
(*ii*) 4.39476	4.395
(*iii*) 3.85231	3.852
(*iv*) 15.6483	15.65
(*v*) 75.384	75.38
(*vi*) 61.555	61.56

Example 5.2. *Let $a = 12.32517$. What is the value of a when*

(*a*) *a is rounded to two places of decimal;*

(*b*) *a is truncated to two places of decimal?*

Also, determine the error in each case.

Solution: Given $a = 12.32517$

(*a*) Rounded value

$$a_1 = 12.33$$

$$\therefore \quad \text{Error} = 12.32517 - 12.33$$

$$= 0.00483 \quad \textbf{Ans.}$$

(*b*) Truncated Value

$$a_1 = 12.32$$

$$\therefore \quad \text{Error} = 12.32517 - 12.32$$

$$= 0.00517 \quad \textbf{Ans.}$$

5.3 RELATIVE ERROR AND NUMBER OF SIGNIFICANT FIGURES

The accuracy of a measurement or a computed result is indicated by the number of significant figures required to express it and not by the number of decimals required. Actually speaking the true index of the accuracy of a measurement or a calculation is only its relative error. For example, if the diameter of a one metre steel shaft is measured to the nearest millimetre, the result is less accurate than the measurement of a kilometre length of a canal to the nearest centimetre. The absolute errors and relative errors in both cases are:

	Absolute error	Relative error
Shaft	0.5 mm	$0.0005 = \frac{1}{4000}$
Canal length	0.5 cm	$\frac{1}{100000}$

Here, we notice that in the measurement of the shaft we made an error of one part in 4000 whereas in the case of the canal, we made an error of one part in 100,000. The measurement of latter is made with a better accuracy as compared to that of the first.

The relation between the relative error and the number of correct figures is given by the following fundamental theorems:

Theorem I. *If the first significant figure of a number is K, and the number is correct to n significant figures, then the relative error is less than* $\frac{1}{(k \times 10^{n-1})}$.

However, the converse of the theorem is not true.

Theorem II. *If the relative error in an approximate number is less than* $\frac{1}{[(k+1) \times 10^{n-1}]}$, *the number is correct to n significant figures, or at least is in error by less than a unit in the nth significant figure.*

It may further be stated that the absolute error is connected with the number of decimal places, whereas the relative error is connected with the number of significant figures.

Application of the error formulae to the fundamental operations of Arithmetic and to logarithms.

1. Addition

We know if $N = u_1 + u_2 + u_3 + \dots + u_n$

then $E_a = \Delta N = \Delta u_1 + \Delta u_2 + \Delta u_3 + \dots + \Delta u_n$

i.e., Absolute error of a sum of approximate numbers is equal to the algebraic sum of their absolute errors.

Example 5.3. *Find the sum of the approximate numbers 459.35, 565.5, 95.953, 3.9244, if each number is only correct to its last figure.*

Solution: In this problem, second number is correct only to the first decimal place. Hence, we should round off the other members to two decimals, one more than that of second, add the four numbers and finally get result to one decimal place by rounding off.

$$\begin{array}{r} 459.35 \\ 565.50 \\ 95.95 \\ 3.92 \\ \hline 1124.72 \\ \hline \end{array}$$

∴ The sum = 1124.7 correct to one decimal place. **Ans.**

Note. It may be noted that rounding off the numbers of more accuracy is done to one more place of decimal as compared to the number of least, accuracy. By doing so the errors inherent to these numbers are eliminated. The result of addition is rounded off to the place of decimal of the least error. However, it may be clearly understood that final result is uncertain by one unit in its last figure.

Example 5.4. *Find the sum of 85930, 826, 526.15, 35200, 7538, 2153, assuming that the number 35200 is known to be correct only to three significant figures.*

Solution: As one of the numbers is only correct to the nearest hundred, we round off the other numbers to the nearest tens, add them and finally give the result to hundreds.

$$\begin{array}{r} 35200 \\ 85930 \\ 7540 \\ 2150 \\ 830 \\ 530 \\ \hline \end{array}$$

$\therefore$ Sum = 132180 **Ans.**

2. Average:

The reliability of the mean of several approximate numbers, is more than that of the numbers from which it was obtained. Consider the following examples:

Let the mantissas of 5 consecutive logarithms taken from a six place table are tabulated in the first column. The same mantissas rounded off to the five place, are tabulated in the second column. The third column gives the errors due to rounding, expressed in units of the sixth decimal place.

	I		II	III
	0.961563		0.96156	3
	0.961611		0.96161	1
	0.961658		0.96166	–2
	0.961706		0.96171	– 4
	0.961753		0.96175	3
Sum =	4.808291	=	4.80829	+ 1
Average =	0.9616582	=	0.961658	Average, 0.2

We have the relation

$$N = N' + E$$

i.e., $$\Sigma N = 2N' + \Sigma E$$

and $\dfrac{\Sigma N}{n} = \dfrac{\Sigma N'}{n} + \dfrac{\Sigma E}{n}$ where n is the number of the approximate numbers.

It may be noticed that the average of the rounded numbers is in error by only a unit in the sixth decimal place. In other words the mean is correct to one more place than the rounded numbers.

Note: The following points may be noted:

(*i*) The entries in all numerical tables are rounded numbers in which the error is not greater than half a unit in the last significant figure.

(*ii*) As the errors introduced due to rounding off are most likely to be positive or negative, their algebraic sum is never large.

You may now understand why the computer retains one more figure in the mean of a set of numbers than are given in the numbers themselves.

3. Subtraction

In this case

$$N = u_1 - u_2$$

and hence $\Delta N = E_a = \Delta u_1 - \Delta u_2$; where Δu_1 and Δu_2 may be either positive or negative. For all cases, the sum of the absolute values of the errors are taken in order to get the maximum error, *i.e., the absolute error of the difference of two approximate numbers may equal the sum of their absolute errors.*

Before subtracting one approximate number from another, they are rounded off to the same place.

Example 5.5. *Subtract 35.674 from 852.4, assuming that each number is approximate and correct only to its last figure.*

Solution: Round the first number to the first place, ***i.e.,*** 35.7

Now difference $= 852.4 - 35.7 = 816.7$

One should safeguard against the following mistake:

$852.400 - 35.674 = 816.726$

It is a wrong result because the last two figures in the larger number are not zeros.

In case of subtraction, both the numbers must be rounded off to the same place. For example,

Let 45.837 and 635.2 are two approximate numbers, correct only to its last figure.

The result of the subtraction is under:

$$635.2 - 45.8 = 589.4$$

It is, however, not advisable to write

$$635.200 - 45.837 = 589.363$$

In this case one should note carefully that the last two figures in the larger number are written by inserting zeros and are not correct.

Loss of significant figures by subtraction: The most serious error connected with the subtraction of approximate numbers arises from the subtraction of numbers which are nearly equal. Study the following examples:

(*i*) $85.238 - 84.565 = 0.673$

(*ii*) $27.834 - 26.972 = 0.862$

(*iii*) 45.351 – 44.241 = 1.110

(*iv*) 37.456 - 36.846 = 0.610

The difference (*i*) is correct to three figures.

The difference (*ii*) is correct to three figures.

The difference (*iii*) is correct to three figures.

The difference (*iv*) is correct to two figures.

The loss of significant figures should be carefully watched in the compusation.

5.4 TYPES OF ERRORS

The errors in the computation of approximate data are of the following types:

(*a*) Absolute error

(*b*) Relative error

(*c*) Percentage error.

(*a*) **Absolute error.** The numerical difference between the true value of the quantity and its approximate value as given or obtained by measurement or computation, is called *absolute error.*

Let the true length of a measured base be 505.35 m.

The measured length be 505.12 m.

∴ Absolute error in the measured length

= true length of the base line – measured length of the base line

= 505.35 – 505.12

= 0.23 m.

(*b*) **Relative error.** The ratio of the absolute error and true value of the quantity, is called *'relative error'*.

For the measured base length referred to above, the relative error

$$= \frac{\text{Absolute error}}{\text{True value}} = \frac{0.23}{505.35} = 4.5513 \times 10^{-4}.$$

(*c*) **Percentage error.** The 100 times the relative error, is called the *percentage error*.

In the above case, the percentage error of the measured length

$$= 100 \times \text{relative error}$$

$$= 100 \times 4.5513 \times 10^{-2}.$$

Note. The following points may be noted:

(*i*) The unit of absolute error is the same as that of the unit used in calculation.

(*ii*) The relative error and percentage errors are independent of the unit of measurement.

(*d*) **Truncation error.** The error caused due to replacing an infinite process by a finite one, is called *truncation error.*

For example, consider the following equation:

$$\sin x = x - \frac{x^3}{3!} + \frac{x^5}{5!} - \frac{x^7}{7!} + \ldots \propto = a$$

If we replace the equation by the following one

$$\sin x = x - \frac{x^3}{3!} + \frac{x^5}{5!} = b$$

Then, the truncation error

$$= a - b.$$

5.5 ESTIMATION OF ERRORS IN THE FUNCTIONS

Let $y = f(x_1, x_2, x_3, \ldots, x_n)$

where $x_1, x_2, x_3, \ldots, x_n$ are variables

Let $\Delta x_1, \Delta x_2, \Delta x_3, \ldots \Delta x_n$ be the respective errors in $x_1, x_2, x_3, \ldots, x_n$.

The absolute error

$$= |\Delta y| \leq \left|\Delta x_1 \cdot \frac{\partial y}{\partial x_1}\right| + \left|\Delta x_2 \cdot \frac{\partial y}{\partial x_2}\right| + \ldots + \left|\Delta x_n \cdot \frac{\partial y}{\partial x_n}\right|$$

The relative error

$$= \left|\frac{\Delta y}{y}\right| \leq \left|\frac{\Delta x_1}{y} \cdot \frac{\partial y}{\partial x_1}\right| + \left|\frac{\Delta x_2}{y} \cdot \frac{\partial y}{\partial x_2}\right| + \ldots + \left|\frac{\Delta x_n}{y} \cdot \frac{\partial y}{\partial x_n}\right|$$

Example 5.6. *In a given equation* $P = \frac{3x^2y}{z^3}$, $x = y = z = I$ *and errors in* x, y, z *are 0.001.*

Calculate the possible maximum error in P.

Solution: We know that

$$\left|\frac{\Delta P}{P}\right| \leq \left|\frac{\Delta x}{P} \cdot \frac{\partial P}{\partial x}\right| + \left|\frac{\Delta y}{P} \cdot \frac{\partial P}{\partial y}\right| + \left|\frac{\Delta z}{P} \cdot \frac{\partial P}{\partial z}\right|$$

$$\leq \left|\frac{\frac{\Delta x}{3x^2y}}{z^3} \cdot \frac{6yx}{z^3}\right| + \left|\frac{\frac{\Delta y}{3x^2y}}{z^3} \cdot \frac{3x^2}{z^3}\right| + \left|\frac{\frac{\Delta z}{3x^2y}}{z^3} \cdot \frac{-9x^2y}{z^4}\right|$$

$$\leq \left|\Delta x \cdot \frac{z^3}{3x^2y} \cdot \frac{6yz}{z^3}\right| + \left|\Delta y \cdot \frac{z^3}{3x^2y} \cdot \frac{3x^2}{z^3}\right| - \left|\Delta z \cdot \frac{z^3}{3x^2y} \cdot \frac{9x^2y}{z^4}\right|$$

$$\leq \left|\frac{2\Delta x}{x}\right| + \left|\frac{\Delta y}{y}\right| - \left|\frac{3\Delta z}{z}\right|$$

$$\leq \frac{2 \times .001}{1} + \frac{0.001}{1} - \frac{3 \times 0.001}{1}$$

$$\leq 0.002 + 0.001 - 0.003$$

$\leq 0.000.$ **Ans.**

5.6 INTERPOLATION

In elementary mathematics, the process of reading in between the lines of a table or computing an intermediate value of a function from a set of given values or tabular values of that function, is called *interpolation.* In higher mathematics, we come across complicated functions whose analytical form is either totally unknown or else is of such a nature that computation cannot easily be subjected to such operations that may be required. In such cases, the given function need be replaced by another function which can be easily handled for computation.

Suppose $y = f(x)$ is a function whose values are $y_0, y_1, y_2, y_3, ..., y_n$ if $x_0, x_1, x_2, x_3, ..., x_n$ are substituted respectively for its independent variable x.

Now, consider another function say $\phi(x)$ which is a simpler function and may take the same values of $x_0, x_1, x_2, x_3, ..., x_n$ for its independent variable x.

Now, the function $f(x)$ may be easily replaced by the function $\phi(x)$ over a given interval. In higher mathematics, the operation of replacing or representing a given furtction by a simpler one, is defined as *interpolation.*

As the function $\phi(x)$ may represent a polynomial, a finite trigonometric series or a series of exponential functions, the process of representing the function $f(x)$ by $\phi(x)$ is called *polynomial interpolation, trigonometric interpolation or exponential interpolation* respectively. In all practical problems, it is convenient to choose for $\phi(x)$, a suitable form of polynomial which are simplest functions and nearly all the standard formulae of interpolation are also polynomial formulae.

1. **Every function which is continuous in an interval (a, b) can be represented in that interval to any desired degree of accuracy by a polynomial.** It means that it is possible to find a polynomial $\phi(x)$ such that $f(x) - \phi(x) | < e$ for every value of x in the interval (a, b) where e is any preassigned positive quantity.

2. **Every continuous function of period 2π can be represented by a finite trigonometrical series of the form**

$$g(x) = a_1 + a_1 \sin x + a_2 \sin 2x + ... + a_n \sin n x + b_1 \cos x + b \cos 2x + ... + b_n \cos nx$$

It means that it is possible to find a finite trigonometric series $g(x)$ such that $|f(x) - g(x)| < \delta$ for all values of x in the interval as considered, where δ is a preassigned positive quantity.

According to these fundamental theorems the given function may be replaced by a polynomial or by a finite trigonometric series to any desired degree of accuracy.

5.7 DIFFERENCES

Consider a function $y = f(x)$ such that $y_0, y_1, y_2, y_3, \dots y_n$ denote a set of its values.

then $y_1 - y_0, y_2 - y_1, y_3 - y_2, y_4 - y_3, \dots, y_n - y_{n-1}$

are called the *first differences* of the function y.

Let

$$\begin{aligned} y_1 - y_0 &= \Delta y_0 \\ y_2 - y_1 &= \Delta y_1 \\ y_3 - y_2 &= \Delta y_2 \\ y_4 - y_3 &= \Delta y_3 \\ &\dots\dots \\ &\dots\dots \\ y_n - y_{n-1} &= \Delta y_{n-1} \\ y_{n+1} - y_n &= \Delta y_n \text{ then} \end{aligned}$$

$$\begin{aligned} \Delta^2 y_0 &= \Delta y_1 - \Delta y_0 = y_2 - y_1 - y_1 + y_0 = y_2 - 2y_1 + y_0 \\ \Delta^2 y_1 &= \Delta y_2 - \Delta y_1 = y_3 - y_2 - y_2 + y_1 = y_3 - 2y_2 + y_1 \\ \Delta^2 y_2 &= \Delta y_3 - \Delta y_2 = y_4 - y_3 - y_3 + y_2 = y_4 - 2y_3 + y_2 \\ &\dots\dots \\ &\dots\dots \\ &\dots\dots \\ \Delta^2 y_n &= \Delta y_{n+1} - \Delta y_n = y_{n+1} - y_n - y_n + y_{n-1} = y_{n-1} - 2y_n + y_{n-1} \end{aligned}$$

$\Delta^2 y_0, \Delta^2 y_1, \Delta^2 y_2, \dots \Delta^2 y_n$ which are differences of the first differences, are called **second differences** of the function y.

Similarly, the third differences are:

$$\begin{aligned} \Delta^3 y_0 &= \Delta^2 y_1 - \Delta^2 y_0 = y_3 - 2y_2 + y_1 - (y_2 - 2y_1 + y_0) \\ &= y_3 - 3y_2 + 3y_1 - y_0 \\ \Delta^3 y_1 &= \Delta^2 y_2 - \Delta^2 y_1 = y_4 - 2y_3 + y_2 - (y_3 - 2y_2 + y_1) \\ &= y_4 - 3y_3 + 3y_2 - y_1 \\ &\dots\dots \\ &\dots\dots \\ \Delta^3 y_{n-1} &= \Delta^2 y_n - \Delta^2 y_{n-1} = y_{n+2} - 3y_{n+1} + 3y_n - y_{n-1} \end{aligned}$$

Differences of a polynomial. Let us now compute the successive differences of a polynomial of the nth degree as under:

$y = f(x) = ax^n + bx^{n-1} + cx^{n-2} + dx^{n-3} \ldots + kx + l$ ...(1)

Replacing $y = (y + \Delta y)$ and $x = (x + \Delta x)$ in eqn. (1), we get

$y + Dy = a(x + \Delta x)^n + b(x + \Delta x)^{n-1} + c(x + \Delta x)^{n-2} + \ldots + k(x + \Delta x) + l$

Substituting $h = \Delta x$ in this equation we get

$y + \Delta y = a(x + h)^n + b(x + h)^{n-1} + c(x + h)^{n-2} + \ldots + k(x + h) + l$ ...(2)

Substituting eqn. (1) from eqn. (2), we get

$\Delta y = a[(x + h)^n - x^n] + b[(x + h)^{n-1} - x^{n-1}] + c[(x + h)^{n-2} - x^{n-2}]$
$+ d[(x + h)^{n-3} - x^{n-3}] + \ldots + k[(x + h) - x],$

i.e., $\Delta y = a[(x + h)^n - x^n] + b[(x + h)^{n-1} - x^{n-1}]$
$+ c[(x + h)^{n-2} - x^{n-2}] + d[(x + h)^{n-3} - x^{n-3}] + \ldots + kh$ ...(3)

Expanding the quantities $(x + h)^n$, $(x + h)^{n-1}$, $(x + h)^{n-2}$, $(x + h)^{n-3}$, etc, by Binomial theorem, we get

$$\Delta y = a\left[x^n + xhx^{n-1} + \frac{n(n-1)}{2!}h^2x^{n-2} + \frac{n(n-1)(n-2)}{3!}h^3x^{n-3} + \ldots x^n\right]$$

$$+ b\left[x^{n-1} + (n-1)hx^{n-2} + \frac{n(n-1)(n-2)h^2x^{n-3}}{2!} + \ldots x^{n-1}\right]$$

$$+ c\left[x^{n-2} + (n-2)hx^{n-3} + \frac{n(n-2)(n-3)h^2x^{n-4}}{2!} + \ldots - x^{n-2}\right] + \ldots + kh$$

or $$\Delta y = anh\, x^{n-1} + \frac{an(n-1)}{2!}h^2\, x^{n-2} + \frac{an(n-1)(n-2)}{3!}h^3\, x^{n-3}$$

$$+ b(n-1)\, hx^{n-2} + \frac{b(n-1)(n-2)\, h^2x^{n-3}}{2!}$$

$$+ c(n-2)\, hx^{n-3} + \frac{c(n-2)(n-3)\, h^2x^{n-4}}{2!}$$

or $$\Delta y = an\, h\, x^{n-1} + \left[\frac{an(n-1)}{2}h^2 + b(n-1)h\right]x^{n-2}$$

$$+ \left[\frac{an(n-1)(n-2)}{3!}h^3 + \frac{b(n-1)(n-2)}{2!}h^2 + c(n-2)h\right]x^{n-3} + \ldots$$

Assuming $\Delta x = h = a$ constant, the bracketed coefficients of x^{n-2}, x^{n-3}, x^{n-4}, ... are also constant.

i.e., $\Delta y = a_n hx^{n-1} + b'x^{n-2} + c'x^{n-3} + \ldots + k'x + l'$ ...(4)

where b', c', ..., k', etc. are new coefficients.

From eqn. (4), it is seen that the first difference of a polynomial of nth degree is another polynomial of $(n-1)$ degree.

Similarly, it can be proved that the second difference of the polynomial of the $(n-1)$th degree is a polynomial of the $(n-2)$ degree.

The further differences, *i.e.*, third, fourth, fifth, ..., nth, are as under.

$$Dy = an\,hx^{n-1} + b'x^{n-2} + c'x^{n-3} + \ldots + k'x + l'$$

$$\Delta^2 y = an(n-1)\,h^2\,x^{n-2} + b''\,x^{n-3} + c''\,x^{n-4} + \ldots + k''\,x + l''$$

$$\Delta^3 y = an(n-1)\,(n-2)\,h^3 x^{n-3} + b'''\,4x^{n-4} + c'''\,x^{n-5} + \ldots + k'''\,x + l'''$$

....

....

....

$$\Delta^n y = a[n(n-1)\,(n-2)\,(n-3)\ldots 1]\,h^n x^{n-n} = a h^n n.$$

i.e., the nth difference of a polynomial is a constant, and all the higher differences are zero.

In other words, the nth differences of a polynomial of the nth degree are constant when the values of the independent variables are taken in arithmetic progression, *i.e.*, at equal intervals.

The converse of this statement is equally true, *i.e.*, if the nth differences of a tabulated function are constant when the values of the independent variable are taken in arithmetic progression, the function is a polynomial of n degree.

From this converse statement, it becomes clear that any function can be replaced by a polynomial if its differences of some order become constant or nearly constant.

5.8 NEWTON'S FORMULAE FOR INTERPOLATION

Newton suggested two formulae for interpolation, *i.e.*,

(1) Formula for Forward interpolation

(2) Formula for Backward interpolation.

(1) **Forward Interpolation.** According to this formula any function may be replaced by a suitable polynomial over a given interval.

Let $y = f(x)$ be the given function.

$y_0, y_1, y_2, \ldots, y_n$ be its values for the equidistant values $x_0, x_1, x_2, \ldots, x_n$ of the independent variable x

$y = \phi(x)$ be a polynomial of the nth degree.

Let the polynomial for the function $y = \phi(x)$ be written as under:

$$\phi(x) = a_0 + a_1(x - x_0) + a_2(x - x_0)\,(x - x_1) + a_3(x - x_0)\,(x - x_1)\,(x - x_2)$$
$$+ a_4(x - x_0)\,(x - x_1)\,(x - x_2)\,(x - x_3) + \ldots$$
$$+ a_n(x - x_0)\,(x - x_1)\,(x - x_2)\ldots(x - x_{n-1}) \qquad \ldots(A)$$

Now, the coefficients $a_0, a_1, a_2, a_3, \ldots a_n$ are determined in such a way that $\phi(x_0) = y_0, (x_1) = y_1, \phi(x_2) = y_2 \ldots \phi(x_n) = y_n$.

Substituting the successive values $x_0, x_1, x_2, x_3 \ldots x_n$ for x in the eqn. (A) and at the sametime putting $\phi(x_0) = y_0, \phi(x_1) = y_1, \phi(x_2) = y_2 \ldots \phi(x_n) = y_n$, we get

$$y_0 = a_0$$

or $$a_0 = y_0$$

$$y_1 = a_0 + a_1(x_1 - x_0)$$

or $$y_1 = y_0 + a_1 h \qquad \text{where } x_1 - x_0 = h$$

$$\therefore \quad a_1 = \frac{y_1 - y_0}{h} = \frac{\Delta y_0}{h}$$

$$y_2 = a_0 + a_0 + a_1(x_2 - x_0) + a_2(x_2 - x_0)(x_2 - x_1)$$

$$= y_0 + \frac{(y_1 - y_0)}{h}(2h) + a_2(2h)(h)$$

where $$x_2 - x_0 = 2h$$

and $$(x_2 - x_1) = h$$

$$y_2 = y_0 + (y_1 - y_0) + a_2 2h^2$$

$$a_2 = (y_2 - y_0 - (2y_1 - 2y_0)\frac{1}{2h^2}$$

$$= (y_2 - y_0 - 2y_1 - 2y_0)\frac{1}{2h^2}$$

$$= \frac{y_2 - 2y_1 + y_0}{2h^2} = \frac{\Delta_2 y_0}{2h^2}$$

$$y_3 = a_0 + a_1(x_3 - x_0) + a_2(x_3 - x_0)(x_3 - x_1) + a_3(x_3 - x_0) + (x_3 - x_0)(x_3 - x_2)$$

$$y_3 = y_0 + \frac{y_1 - y_0}{h}(3h) + \frac{y_2 - 2y_1 + y_0}{2h^2}(3h)(2h) + a_3(3h)(2h)(h)$$

or $$y_3 = y_0 + 3(y_1 - y_0) + (y_2 \; 2y_1)^3 + a_3 \, 6h^3$$

$$\therefore \quad a_3 = [y_3 - y_0 - 3(y_1 - y_0) - 3(y_2 - 2y_1 + y_0] \frac{1}{6h^3}$$

$$[y_3 - y_0 - 3y_1 + 3y_0 - 3y_2 + 6y_1 - 3y_0] \frac{1}{6h^3}$$

$$\therefore \quad a_3 = (y_3 - y_0 - 3y_1 + 3y_0 - 3y_2 + 6y_1 - 3y_0) \frac{1}{6h^3}$$

$$= \frac{y_3 - 3y_2 + 3y_1 - y_0}{6h^3} = \frac{\Delta_3 y_0}{3!\,h^3}$$

$$y_4 = a_0 + a_1(x_4 - x_0) + a_2(x_4 - x_0)(x_4 - x_1) + a_3(x_4 - x_0)(x_4 - x_1)(x_4 - x_2) + a_4(x_4 - x_0)(x_4 - x_1)(x_4 - x_2)(x_4 - x_3)$$

or
$$y_4 = y_0 + \frac{y_1 - y_0}{h}(4h) + \frac{y_2 - 2y_1 + y_0}{2h^2}(4h)(3h) + \frac{y_3 - 3y_2 + 3y_1 - y_0}{6h^3}(4h)(3h)(2h) + a_4(4h)(3h)(2h)(h)$$

or
$$y_4 = y_0 - (y_1 - y_0) + (y_2 - 2y_1 + y_0)6 + (y_3 - 3y_2 + 3y_1 - y_0)4 + a_4\, 24h^4$$

$$a_4 = \frac{(y_4 - y_0 - 4y_1 + 4y_0 - 6y_2 + 12y_1 - 6y_0 - 4y_3 + 12y_2 - 12y_1 + 4y_0}{24h^4}$$

or
$$a_4 = \frac{y_4 - 4y_3 - 6y_2 - 4y_1 + y_0}{24h^4} = \frac{\Delta_4 y_0}{4!\,h^4}$$

....

....

$$a_5 = \frac{\Delta_5 y_0}{5!\,h^5}$$

$$a_6 = \frac{\Delta_6 y_0}{6!\,h^6}$$

$$a_7 = \frac{\Delta_7 y_0}{7!\,h^7}$$

....

....

$$a_n = \frac{\Delta_n y_0}{n!\,h^n}$$

Substituting the values of $a_0, a_1, a_2, a_3 \ldots a_n$ as obtained above, in eqn. (A), we get

$$\phi(x) = y_0 + \frac{\Delta y_0}{h}(x - x_0) + \frac{\Delta_2 y_2}{2!\,h^2}(x - x_0)(x - x_1)$$

$$+ \frac{\Delta_3 y_0}{3!\,h^3}(x - x_0)(x - x_1)(x - x_2) + \frac{\Delta_4 y_0}{3!\,h^4}(x - x_0)(x - x_1)(x - x_2)(x - x_3) + \ldots$$

$$+ \frac{\Delta_n y_0}{n!\,h^4}(x - x_0)(x - x_1)(x - x_2)\ldots(x - x_{n-1})$$

The above equation may further be simplified by a change of variable

$$\phi(x) = y_0 + \Delta y_0 \frac{(x - x_0)}{h} + \frac{\Delta_2 y_0}{2}\left(\frac{x - x_0}{h}\right)\left(\frac{x - x_1}{h}\right)$$

$$+\frac{\Delta_3 y_0}{3!}\left(\frac{x-x_0}{h}\right)\left(\frac{x-x_1}{h}\right)\left(\frac{x-x_2}{h}\right)$$

$$+\frac{\Delta_4 y_0}{4!}\left(\frac{x-x_0}{h}\right)\left(\frac{x-x_1}{h}\right)\left(\frac{x-x_2}{h}\right)\left(\frac{x-x_3}{h}\right) \quad \text{...(B)}$$

Now, put $\frac{x-x_0}{h} = u$ or $x = x_0 + hu$

We know that

$$x_1 = x_0 + h;\ x_2 = x_0 + 2h;\ x_3 = x_0 + 3h,\ ...$$

$$\therefore \quad \frac{x-x_1}{h} = \frac{x-(x_0+h)}{h} = \frac{x-x_0-h}{h} = \frac{x-x_0}{h} - 1 = u-1$$

$$\frac{x-x_2}{h} = \frac{x-(x_0+2h)}{h} = \frac{x-x_0-2h}{h} = \frac{x-x_0}{h} - 2 = u-2$$

$$\frac{x-x_3}{n} = \frac{x-(x_0+3h)}{h} = \frac{x-x_0-3h}{h} = \frac{x-x_0}{h} - 3 = u-3$$

....

....

$$\frac{x-x_{n-1}}{h} = \frac{x-[x_0+(n-1)h]}{h} = \frac{x-x_0-(n-1)h}{h}$$

$$= \frac{x-x_0}{h} - (n-1) = u-(n-1)$$

Substituting the values of $\frac{x-x_0}{h}, \frac{x-x_1}{h}, \frac{x-x_2}{h}, \frac{x-x_3}{h}, ..., \frac{x-x_{n-1}}{h}$ in equation (B), we get

$$\phi(x) = \phi(x_0 + hu) = \phi(u)$$

$$= y_0 + u\Delta y_0 + \frac{u(u-1)}{2!}\Delta^2 y_0 + \frac{u(u-1)(u-2)}{3!}\Delta^3 y_0$$

$$+ \frac{u(u-1)(u-2)(u-3)}{4!}\Delta^4 y_0$$

....

....

....

$$+ \frac{u(u-1)(u-2)(u-3)\cdots(u-n+1)}{n!}\Delta^n y_0 \quad \text{...(C)}$$

This equation whose coefficients of Δ's are the binomial coefficients, is the standard Newton's formula or *forward interpolation.*

Note. The following points may be noted:

(*i*) The formula is said to be forward interpolation because the values of the tabulated functions are from y_0 onward to right.

(*ii*) The formula is used for interpolating the values of y near the beginning of the tabulation.

(*iiii*) The starting point may be chosen at any tabulated value but thereafter values are taken to the right.

(2) **Backward Interpolation.** The forward interpolation formula can not be used for interpolating a value of y near the end of the tabular data. To derive a suitable formula for interpolating the data at the end of the tabulated values, consider the polynomial of the following type.

$$\begin{aligned}\phi(x) &= a_0 + a_1(x - x_n) + a_2(x - x_n)(x - x_{n-1}) \\ &\quad + a_3(x - x_n)(x - x_{n-1})(x - x_{n-2}) \\ &\quad + a_n(x - x_n)(x - x_{n-1})(\ldots)(x - x_1) \quad \ldots(A)\end{aligned}$$

Now, we have to determine the values of coefficients $a_0, a_1, a_2, a_3, a_4, a_5 \ldots a_n$ so that

$$\begin{aligned}\phi(x_n) &= y_n \\ f(x_{n-1}) &= y_{n-1} \\ f(x_{n-2}) &= y_{n-2} \\ f(x_{n-3}) &= y_{n-3} \\ &\ldots\ldots\ldots\ldots \\ &\ldots\ldots\ldots\ldots\end{aligned}$$

Substituting the values of $x_n, x_{n-1}, x_{n-2} \ldots$

for x and putting $\phi(x_n) = y_n, \phi(x_{n-1}) = y_{n-1} \ldots$, we have

$$y_n = a_0$$

or $$a_0 = y_n$$

$$\begin{aligned}Y_{n-1} &= a_0 + a_1(x_{n-1} - x_n) \\ &= y_n + a_1(-h)\end{aligned}$$

$$[\because \; x_n - x_{n-1} = h \text{ or } x_{n-1} - x_n = -h]$$

$\therefore$ $$a_1 = \frac{y_n - y_{n-1}}{h} = \frac{\Delta_1 y_n}{h}$$

$$\begin{aligned}y_{n-2} &= a_0 + a_1(x_{n-2} - x_n) + a_2(x_{n-2} - x_n)(x_{n-2} - x_{n-1}) \\ &= y_n + \frac{y_n - y_{n-1}}{h}(-2h) + a_2(-2h)(-h)\end{aligned}$$

or $$y_{n-2} = y_n - 2(y_n - y_{n-1}) + 2a_2h^2$$

$$\therefore \quad a_2 = \frac{y_{n-2} - y_n + 2y_n - 2y_{n-1}}{2h^2} = \frac{y_n - 2y_{n-1} + y_{n-2}}{2h^2}$$

$$= \frac{\Delta_2 y_n}{2h^2}$$

$$y_{n-3} = a_0 + a_1(x_{n-3} - x_n) + a_2(x_{n-3} - x_n)(x_{n-3} - x_{n-1}) + a_2(x_{n-3} - x_n)(x_{n-3} - x_{n-1})(x_{n-3} - x_{n-2})$$

$$y_{n-3} = \frac{y_n - y_{n-1}}{h}(-3h) + \frac{y_n - 2y_{n-1} + y_{n-2}}{2h^2}(-2h)(-2h) + a_3(-3h)(-2h)(-h)$$

$$y_{n-3} = y_n - 3(y_n - y_{n-1}) + 3(y_n - 2y_{n-1} + y_{n-2}) - a_3 6h^3$$

$$a_3 = \frac{y_n - y_{n-3} - 3y_{n-1} + 3y_n - 6y_{n-1} + 3y_{n-2}}{6h^3}$$

$$= \frac{y_n - 3y_{n-1} + 3y_{n-2} - y_{n-3}}{3!\,h^3} = \frac{\Delta_3 y_n}{3!\,h^3}$$

Similarly we may get

$$a_4 = \frac{\Delta_4 y_n}{4!\,h^4},\ a_5 = \frac{\Delta_4 y_n}{5!\,h^5},\ \ldots,\ a_n = \frac{\Delta_n y_n}{n!\,h^n}$$

Now, substitute the values of a_0, a_1, a_3, ..., a_n in eqn. (A) to get

$$\phi(x) = y_n + \frac{\Delta_1 y_n}{h^4}(x - x_n) + \frac{\Delta_2 y_n}{2h}(x - x_n)(x - x_{n-1})$$

$$+ \frac{\Delta_3 y_n}{3!\,h^3}(x - x_n)(x - x_{n-1})(x - x_{n-2})$$

$$+ \frac{\Delta_4 y_n}{4!\,h^4}(x - x_n)(x - x_{n-1})(x - x_{n-2})(x - x_{n-3}) + \ldots$$

$$+ \frac{\Delta_n y_n}{n!\,h^n}(x - x_n)(x - x_{n-1})(x - x_{n-2})(x - x_{n-3}) + (x - x_0)$$

Again, the above equation may be written as

$$\phi(x) = y_n + \Delta_1 y_n\left(\frac{x - x_n}{h}\right) + \frac{\Delta_2 y_n}{2}\left(\frac{x - x_n}{h}\right)\left(\frac{x - x_{n-1}}{h}\right)$$

$$+ \frac{\Delta_3 y_n}{3!}\left(\frac{x - x_n}{h}\right)\left(\frac{x - x_{n-1}}{h}\right)\left(\frac{x - x_{n-2}}{h}\right)$$

$$+ \frac{\Delta_4 y_n}{4!}\left(\frac{x - x_n}{h}\right)\left(\frac{x - x_{n-1}}{h}\right)\left(\frac{x - x_{n-2}}{h}\right)\left(\frac{x - x_{n-3}}{h}\right)$$

....

....

....

$$+\frac{\Delta_n y_n}{n!}\left(\frac{x-x_n}{h}\right)\left(\frac{x-x_{n-1}}{h}\right)\left(\frac{x-x_{n-2}}{h}\right)\cdots\left(\frac{x-x_1}{h}\right) \quad \text{...(C)}$$

Putting $u=\left(\frac{x-x_n}{h}\right)$ or $x=x_n+hu$

and remembering that

$$x_{n-1} = x_{n-h};\ x_{n-2}=x_n-2h, \text{ etc., we get}$$

$$\frac{x-x_{n1}}{h} = \frac{x-(x_n-h)}{h}=\frac{x-x_n+h}{h}=\frac{x-x_n}{h}+1=u+1$$

$$\frac{x-x_{n-2}}{h} = \frac{x-(x_n-2h)}{h}=\frac{x-x_n+2h}{h}=\frac{x-x_n}{h}+2=u+2$$

$$\frac{x-x_{n-3}}{h} = \frac{x-(x_n-3h)}{h}=\frac{x-x_n+3h}{h}=\frac{x-x_n}{h}+3=u+3$$

....

....

....

$$\frac{x-x_1}{h} = \frac{x-(x_n-(n-1)h)}{h}=\frac{x-x_n}{h}=\frac{x-1}{h}h=u+(n-1)$$

Substituting the values of $\frac{x-x_n}{h}, \frac{x-x_{n-1}}{h}, \frac{x-x_{n-2}}{h},$

..., $\frac{x-x_1}{h}$ in eqn. (C), we get

$$\phi(x) = \phi(x_n+h)=\phi(u)$$

$$= y_n+u\Delta_1 y_n+\frac{u(u+1)}{2}\Delta_2 y_n+\frac{u(u+1)(u+2)}{3!}\Delta_3 y_n$$

$$+\frac{u(u+1)(u+2)(u+3)}{4!}\Delta_4 y_n+\ldots$$

$$+\frac{u(u+1)(u+2)(u+3)(u+4)\ldots(u+n-1)}{n!}\Delta_n y_{n-1}$$

This formula is most useful for interpolating the values of y near the end of a set of tabular values.

Example 5.7. *Expande $f(x) = \cos x$ into a Maclaurin series and compute the value of cos 0.2 to four decimal places.*

Solution: We know by Maclaurin's theorem

$$f(x) = f(0) + xf'(0) + \frac{x_2}{2!}f''(0) + \frac{x_3}{3!}f^2(0) + \ldots \qquad \ldots(i)$$

Here,

$$\begin{aligned} f(x) &= \cos x & f(0) &= \cos 0 = 1 \\ f'(x) &= -\sin x & f'(0) &= -\sin 0 = 0 \\ f''(x) &= -\cos x & f'(0) &= -\cos 0 = -1 \\ f'''(x) &= \sin x & f'''(0) &= \sin 0 = 0 \\ f'^{v}(x) &= \cos x & f^{iv}(0) &= \cos 0 = 1 \end{aligned}$$

....

....

....

and so on.

Substituting these values in eqn. (*i*), we get

$$\cos x = 1 \times x \times 0 + \frac{x^2}{2} \times (-1) + \frac{x^3}{6} \times 0 + \frac{x^4}{24} \times (1) + \ldots$$

$$\cos x = 1 - \frac{x^2}{2} + \frac{x^4}{24} - \ldots \qquad \ldots(ii)$$

Putting $x = 0.2$ in eqn. (*ii*), we get

$$\cos(0.2) = 1 - \frac{(0.2)^2}{2} + \frac{(0.2)^4}{24}$$

$$= 1 - \frac{0.04}{2} + \frac{0.0016}{24}$$

$$= 1 - 0.02 + 0.000667$$

$$= 1 - 0.02 + 0.0007$$

$$\therefore \quad \cos(0.2) = 0.9807$$

5.9 NUMERICAL INTEGRATION

The process of computing the value of a definite integral from a set of numerical values of the integrand is called *numerical integration.* The numerical integration may be obtained by representing the integrand by an interpolation formula usually one involving the differences and then integrating the formula between the desired limits.

Let us integrate the Newton's formula of forward interpolating over n equidistant intervals of width $h = \Delta$, *i.e.*,

$$\int_{0}^{x_0+nh} ydx = h\int_0^n \left(y_0 + u\Delta y_0 + \frac{u(u-1)}{2!}\Delta_2 y_0 + \frac{u(u-1)(u-2)}{3!}\Delta_3 y_0 + \frac{u(u-1)(u-2)(u-3)}{4!}\Delta_4 y_0 + \frac{u(u-1)(u-2)(u-3)(u-4)}{5!}\Delta_5 y_0 + \frac{u(u-1)(u-2)(u-3)(u-4)(u-5)}{6!}\Delta_5 y_0 + \ldots\right) du$$

or
$$\int_{x_0}^{x_0+nh} ydx = h\left[ny_0 + \frac{n^2}{2}\Delta_{y_0} + \left(\frac{n^3}{3} - \frac{n^2}{2}\right)\frac{\Delta^2 y_0}{2} + \left(\frac{n^4}{4} - n^3 + n^2\right)\frac{\Delta^3 y_0}{3!} + \left(\frac{n^5}{5} - \frac{3n^4}{2} + \frac{11n^3}{3} - 3n^2\right)\frac{\Delta^4 y_0}{4!} + \left(\frac{n^6}{6} - 2n^5 + \frac{35n^4}{4} - \frac{50n^3}{3} + 12n^2\right)\frac{\Delta^5 y_0}{5!} + \left(\frac{n^7}{7} - \frac{15n^6}{6} + 17n^5 - \frac{225n^4}{4} - \frac{274n^3}{3} - 60n^2\right)\frac{\Delta^6 y_0}{6!}\right] \quad \ldots(A)$$

From the general formula (A) we obtain a veriety of quadrature formulae by putting $n = 1, 2, 3, 4$, etc. The most suitable and important ones are obtained by putting $n = 2$ and $n = 6$.

1. Simpson's Rule for quadrature

Putting $n = 2$ in eqn. (A) and neglecting all differences above the second, we get

$$\int_{x_0}^{x_0+2h} ydx = h\left[2y_0 + 2\Delta^1 y_0 + \left(\frac{8}{3} - 2\right)\frac{\Delta^2 y_0}{2}\right]$$

$$= h\left[2y_0 + 2y_1 - 2y_0 + \frac{1}{3}(y_2 - 2y_1 + y_0)\right]$$

where $\Delta^1 y_0 = (y_1 - y_0)$ and $\Delta^2 y_0 = y_2 - 2y_1 + y_0$

$$= \frac{h}{3}(y_0 + 4y_1 + y_2) \quad \ldots(i)$$

For the next two intervals, *i.e.*, x_4 to $x_2 + 2h$, we get

$$\int_{x_4}^{x_4+2h} ydx = \frac{h}{3}(y_2 + 4y_3 + y_4) \qquad ...(ii)$$

For the next two intervals, *i.e.*, x_4 to $x_4 + 2h$, we get

$$\int_{x_4}^{x^4+2h} ydx = \frac{h}{3}(y_4 + 4y_5 + y_6) \qquad ...(iii)$$

....

....

....

Adding eqns. (*i*), (*ii*) (*iii*) ... etc., we get

$$\int_{x_0}^{x_0+nh} ydx = \frac{h}{3}(y_0 + 4y_1 + y_2 + y_2 + 4y_3 + y_4$$

$$+ \ldots + y_4 + y_5 + y_6 + \ldots + y_n)$$

or $$\int_{x_0}^{x_0+nh} ydx = \frac{h}{3}[y_0 + 4(y_1 + y_3 + y_5 + \ldots) + 2(y_2 + y_4 + y_6 + \ldots) + y_n]$$

$$= \frac{h}{3}\sum_0^n cy$$

where $c = 1, 4, 2, \ldots 2, 4, 1$

This formula is popularly known as *Simpson's rule.*

It may be clearly noted that the interval of integration must be divided into an even number of sub-intervals of width h.

2. Weddle's Rule for quadrature

Here, we put $n = 6$ in eqn. (A) and neglect all the differences above the sixth.

$$\int_{x_0}^{x_0+6h} ydx = h\left[6y_0 + 18\Delta y_0 + 27\Delta^2 y_0 + 24\Delta^3 y_0 \right.$$

$$\left. + \frac{123}{10}\Delta^4 y_0 + \frac{33}{10}\Delta^5 y_0 + \frac{41}{140}\Delta^6 y_0\right]$$

For the sake of convenience we replace the sixth term $\frac{41}{140}\Delta^6 y_0$ by $\left(\frac{3}{10}\right)\Delta^6 y_0$ without introducing any serious error.

i.e. $$\int_{x_0}^{x_0+6h} ydx = h\left[6y_0 + 18\Delta y_0 + 27\Delta^2 y_0 + 24\Delta^3 y_0 \right.$$

$$\left. + \frac{123}{10}\Delta^4 y_0 + \frac{33}{10}\Delta^5 y_0 + \frac{3}{10}\Delta^6 y_0\right]$$

$$= \frac{3h}{10}[20y_0 + 60\Delta y_0 + 90\Delta^2 y_0 + 80\Delta^3 y_0 + 410\Delta^4 y_0 + 110\Delta^5 y_0 + \Delta^6 y_0]$$

Substituting the values of $y_0 = \Delta y_0, \Delta^2 y_0, \Delta^3 y_0, \Delta^4 y_0, \Delta^5 y_0, \Delta^6 y_0$, we get on simplification

$$\int_{x_0}^{x_0+6h} ydx = \frac{3h}{10}[y_0 + 5y_1 + y_2 + 6y_3 + y_4 + 5y_5 + y_6]$$

For the next six intervals, *i.e.*, from x_6 to x_{12}, we get

$$\int_{x_6}^{x12} ydx = \frac{3h}{10}[y_0 + 5y_7 + y_8 + 6y_9 + y_{10} + 5y_{11} + y_{12}]$$

Adding these expressions, we get

$$\int_{x_0}^{x_0+nh} ydx = \frac{3h}{10}[y_0 + 5y_1 + y_2 + 6y_3 + y_4 + 5y_5 + 2y_6 + 5y_7 + y_8]$$

$$+ 6y_9 + y_{10} + 5y_{11} + 2y_{12} + \ldots + 2y_{n-6} + 5y_{n-5}$$

$$+ y_{n-5} + 6y_{n-3} + y_{n-2} + 5y_{n-1} + y_n]$$

$$= \frac{3h}{10}\sum_{0}^{n} \text{K}y, \text{ where}$$

$$\text{K} = 1, 5, 1, 6, 1, 5, 2, 5, 1, 6, 1, 5, 2 \text{ etc.}$$

This formula to known as *Weddle's Rule* which is more accurate than the Simpson's rule.

3. The Trapezoidal Rule. This rule is applicable to any number of sub-intervals whether even or odd in contrast with Simpson's or Weddle's rules which are only applicable if the interval of integration is divided into an even number of sub-intervals.

The Trapezoidal formula may be derived by adding the areas of all trapezoids between $x = a$ and $x = b$, and by putting $n = 1$ in the following equation

$$\int_{x_0}^{x_0+h} ydx = h\left[ny_0 + \frac{n^2}{2}\Delta y_0 + \left(\frac{n^3}{3} - \frac{n^2}{2}\right)\frac{\Delta^2 y_0}{2}\right.$$

$$+ \left(\frac{n^4}{4} - n^3 + n^2\right)\frac{\Delta^3 y_0}{3!} + \left(\frac{n^5}{5} - \frac{3n^4}{2} + \frac{11n^3}{3} - 3n^2\right)\frac{\Delta^4 y_0}{4!}$$

$$\left. + \left(\frac{n^6}{6} - 2n^5 + \frac{35n^4}{4} - \frac{50n^3}{3} + 12n^2\right)\frac{\Delta^5 y_0}{5!} + (\ldots)\right]$$

Now, evaluate over single sub intervals using the first two terms in the right hand side of the equation. The higher differences than the first are dropped because they involve more than two functional values.

$$\therefore \quad \int_{x_0}^{x_0+h} ydx = h\left(y_0 + \frac{1}{2}\Delta y_0\right) = h\left(y_0 + \frac{y_0 - y_1}{2}\right) = h\left(\frac{y_0 + y_1}{2}\right)$$

$$\int_{x_0}^{x_0+h} ydx = h\left(y_1 + \frac{1}{2}\Delta y_1\right) = h\left(y_1 + \frac{y_2 - y_1}{2}\right) = h\left(\frac{y_1 + y_2}{2}\right)$$

$$\int_{x_0}^{x_0+h} ydx = h\left(y_2 + \frac{y_3 - y_2}{2}\right) = h\left(\frac{y_2 + y_3}{2}\right), \ldots$$

Adding these integrads, we get

$$\int_a^b ydx = h\left(\frac{y_0 + y_1}{2} + \frac{y_1 + y_2}{2} + \frac{y_2 + y_3}{2} + \ldots + \frac{y_{n-1} + y_n}{2}\right)$$

$$= h\left(\frac{1}{2}y_0 + y_1 + y_2 + y_3 + \ldots + y_{n-1} + \frac{1}{2}y_0\right)$$

Note: The following points may be noted:

(*i*) This is the most easy formula.

(*ii*) This is least accurate formula.

(*iiii*) The accuracy may be improved by decreasing the interval h.

Example 5.8.

x	$f(x)$	x	$f(x)$
0.0	1	4.0	5.2
1.0	2.5	5.0	6.0
2.0	4.6	6.0	6.6
3.0	5.0		

From the above table find the integral value of $\int_0^6 f(x)\,dx$ *by using Simpson's rule.*

Solution: By Simpson's rule

$$\int_0^6 f(x)\,dx = \frac{1}{3}[(1 + 6.6) + 2(4.6 + 5.2) + 4(2.5 + 5.0 + 6.0)]$$

$$= \frac{1}{3}[7.6 + 2 \times 9.8 + 4 \times 13.5]$$

$$= \frac{1}{3}[7.6 + 19.6 + 54.0]$$

$$= \frac{1}{3} \times 81.2 = 27.067 \quad \textbf{Ans.}$$

Example 5.9. *Find the value of* $\log_e 2$ *by evaluating the integral* $\int_0^1 \frac{dx}{1+x}$ *numerically using (i) Trapezoidal rule, (ii) Simpson's* $\frac{1}{3}$ *rule, and 10 equal intervals in each case. Compare the results with the value of* $\log_e 2$ *obtained from tables or analytically.*

Solution:

$$\int_0^1 \frac{dx}{1+x} = [\log_e (1 + x)]^1 = \log_e 2$$

$$\int(x) = \frac{1}{1+x}$$

The following is the table of values of x and $f(x)$

x	0	0.1	0.2	0.3	0.4	0.5	0.6	0.7	0.8	0.9	1.0
$fx = \frac{1}{1+x}$	1	0.909	0.833	0.769	0.714	0.667	0.625	0.588	0.556	0.526	0.500

(1) Here $h = 0.1$. Applying Simpson's rule, we get

$$\int_0^1 \frac{dx}{1+x} = \frac{h}{3}[(y_1 + y_{11}) + 2(y_3 + y_5 + y_7 + y_9) + 4(y_2 + y_4 + y_6 + y_8 + y_{10}]$$

$$= \frac{0.1}{3}[(1.000 + 0.5000) + 2(0.833 + 0.714 + 0.625 + 0.556)$$

$$+ 4(0.909 + 0.769 + 0.667 + 0.588 + 0.526]$$

$\therefore \quad \log_e 2 = 0.6931$

(2) $h = 0.1$, Applying the Trapezoidal rule

$$\int_0^1 \frac{dx}{1+x} = h\left(\frac{1}{2}y_0 + y_1 + y_2 + y_3 + y_4 + y_5 + y_6 + y_7 + y_8 + y_9 + \frac{1}{2}y_{10}\right)$$

$$= 0.1(0.5 + 0.909 + 0.833 + 0.769 + 0.714 + 0.667 + 0.625$$

$$+ 0.588 + 0.556 + 0.526 + 0.25)$$

$$= 0.1 \times 6.937 = 0.6937$$

$\therefore \quad \log_e 2 = 0.6937$

From the log table $\log_e 2 = 0.6932539$

Difference from Simpson's rule = 0.6932539 – 0.6931 = 0.0001539.

Difference from Trapezoidal rule = 0.6932539 – 0.6937 = – 0.0004461.

Ans.

Example 5.10. *Compute by any method the value of the integral*

$$\int_0^{\pi/3} \sqrt{\cos\theta}\, d\theta$$

Solution: We shall solve this problem by well known Simpsons $\frac{1}{3}$rd rule.

Let $\int_0^{\pi/3} \sqrt{\cos\theta}\, d\theta = \int_a^b f(\theta)\, d\theta$

$$\therefore \quad a = 0;\ b = \frac{\pi}{3},\ f(\theta) = \sqrt{\cos\theta}$$

Divide the interval $(b - a)$ into 4 equal parts, each of length h.

$$\therefore \quad h = \frac{b-a}{4} = \frac{\frac{\pi}{3} - 0}{4} = \frac{\pi}{12}$$

θ	0	$\frac{\pi}{12}$	$\frac{\pi}{6}$	$\frac{\pi}{4}$	$\frac{\pi}{3}$
$y = f(\theta) = \sqrt{\cos\theta}$	1	0.9828	0.9306	0.8409	0.7071
y	y_1	y_2	y_3	y_4	y_5

Substituting the values in Simpson's $\frac{1}{3}$rd rule, we get

$$\int_a^b f(\theta)\, d\theta = \frac{h}{3}[(y_1 + y_5) + 4(y_2 + y_4) + 2y_3]$$

$$\int_0^{\pi/3} \sqrt{\cos\theta}\, d\theta = \frac{\pi}{36}[(1 + 0.7071) + 4(0.9828 + 0.8409) + 2 \times 0.9306]$$

$$= \frac{\pi}{36}[1.7071 + 7.2948 + 1.8612] = \frac{3.14}{36} \times 10.8631$$

$$= 0.9475 \quad \textbf{Ans.}$$

Example 5.11. *Evaluate using Simpson's rule :*

$I = \int_0^2 \frac{1}{1+x} dx$ *correct to three decimal place. (Take h = 0.125).*

Solution: Given $h = 0.125$

$$x_0 = 0,\ x_1 = 0.125,\ x_2 = 0.250,\ x_3 = 0.375,\ x_4 = 0.5,\ x_5 = 0.625$$

$$x_6 = 0.750,\ x_7 = 0.875,\ x_8 = 1.000,\ x_9 = 1.125,\ x_{10} = 1.250,\ x_{11} = 1.375$$

$$x_{12} = 1.500,\ x_{13} = 1.625,\ x_{14} = 1.750,\ x_{15} = 1.875,\ x_{16} = 2.000$$

i	x_i	$y_i = \frac{1}{1+x}$
0	0	$\frac{1}{1+0} = 1.00000$
1	0.125	$\frac{1}{1+0.125} = 0.88888$
2	0.250	$\frac{1}{1+0.250} = 0.80000$
3	0.375	$\frac{1}{1+0.375} = 0.72727$
4	0.500	$\frac{1}{1+0.5} = 0.64667$
5	0.625	$\frac{1}{1+0625} = 0.61538$
6	0.750	$\frac{1}{1+0.750} = 0.57143$
7	0.875	$\frac{1}{1+0.875} = 0.53333$
8	1.000	$\frac{1}{1+1} = 0.500$
9	1.125	$\frac{1}{1+1.125} = 0.47059$
10	1.250	$\frac{1}{1+1.250} = 0.44444$
11	1.375	$\frac{1}{1+1.375} = 0.42105$
12	1.500	$\frac{1}{1+1.50} = 0.40000$
13	1.625	$\frac{1}{1+1.625} = 0.38095$
14	1.750	$\frac{1}{1+1.750} = 0.36364$
15	1.875	$\frac{1}{1+1.875} = 0.34783$
16	2.000	$\frac{1}{1+2} = 0.33333$

$$\int_0^2 \frac{dx}{1+3} = \frac{h}{3}[y_0 + 4(y_1 + y_3 \ldots) + 2(y_2 + y_4 + \ldots)\, y_{16}]$$

$$= \frac{0.125}{3}[1 + 4(0.88888 + 0.72727 + 0.61538$$

$+ 0.53333 + 0.47059 + 0.42105 + 0.38095 + 0.34783) + 2(0.80000 + 0.66667 + 0.57143 + 0.50000 + 0.44444 + 0.40000 + 0.36364 + 0.33333]$

$$= \frac{0.125}{3}[1 + 17.54112 + 7.49236 + 0.33333]$$

$$= 1.0987617 \quad \textbf{Ans.}$$

Check: $\int_0^2 \frac{dx}{1+x} = [\log(1+x)]_0^2 = \log_{10} 3$

$$\log_{10} 3 = \frac{1}{\log_{10} e} \times \log_e 3 \text{ where } e = 2.7183$$

$$= 1.0987049 \quad \textbf{O.K.}$$

Example 5.12. *Integrate numerically* $\int_0^1 \frac{dx}{1+x^2}$ *using simpson's rule. Take the step-size h as 0.1.*

Solution: Let $I = \int_0^1 \frac{dx}{1+x^2} dx$

Here $f(x) = \frac{dx}{1+x^2}; h = 0.1$

Applying simpson's $\frac{1}{3}$rd rule we get

$$\int_0^1 f(x)dx = \frac{h}{3}[y_1 + y_{n+1}] + 4(y_2 + y_4 + \ldots + y_n) + 2(y_3 + y_5 + \ldots + y_{n-1})] \quad \ldots(i)$$

x	0	0.1	0.2	0.3	0.4	0.5	0.6	0.7	0.8	0.9	1.0
$f(x)$	1	0.9900	0.9615	0.9174	0.8621	0.8000	0.7353	0.6711	0.6098	0.5525	0.5000
y	y_1	y_2	y_3	y_4	y_5	y_6	y_7	y_8	y_9	y_{10}	y_{11}

Substituting the values of y_1 to y_{11} in eqn. (*i*) we get

Substituting the values of y_1 to y_{11} in eqn. (*i*) we get

$$\int_0^1 \frac{1}{1+x^2}\,dx = \frac{0.1}{3}[(1 + 0.5000) + 4(0.9900 + 0.9174 + 0.8000 + 0.6711 + 0.5525) + 2(0.9615 + 0.8621 + 0.7353 + 0.6098]$$

$$= \frac{0.1}{3}[1.5000 + 15.724 + 6.3374]$$

$$= \frac{0.1 \times 23.5614}{3}$$

$$= 0.7854 \quad \textbf{Ans.}$$

Example 5.13. *Calculate the value of* $\int_{-3}^{3} x^4\,dx$ *Simpson's one-third rule (use seven equidistant ordinates). Compare the result with its exact value.*

Solution: Given: 7 equidistant ordinates, *i.e.*, – 3, – 2, – 1, 0, 1, 2, 3 with $y_0, y_1, y_2, y_3, y_4, y_5, y_6$

$$h = 1, \text{ the given interval}$$

The value of integration by Simpson's rule is obtained by the following formula:

$$I = \frac{h}{3}\left[y_0 + 4\frac{(y_1 + y_3 + y_5)}{4} + 2(y_2 + y_4) + y_6\right]$$

$$= \frac{1}{3}[(-3)^4 + 4((-2)^4 + (0)^4 + (2)^4 + 2((-1)^4 + (1)^4) + (3)^4]$$

$$= \frac{1}{3}[81 + 4(16 + 0 + 16) + 2(1 + 1) + 81]$$

$$= \frac{1}{3}[81 + 64 + 0 + 64 + 2 + 2 + 81]$$

$$= \frac{1}{3}[294] = 98$$

The exact result is given by direct integration is

$$\int_{-3}^{3} x^4 dx = \left[\frac{x^5}{5}\right]_{-3}^{3} = \frac{3^5}{5} + \frac{3^5}{5} = \frac{486}{5} = 97.2$$

$$\text{Error} = \text{True value} - \text{observed value}$$

$$= 97.2 - 98 = -0.8$$

∴ The error in Simpson's result = – 0.8. **Ans.**

5.10 ALGEBRAIC AND TRANSCENDENTAL EQUATIONS

The equations of the form $f(x) = 0$ where $f(x)$ is purely a polynomial in x, are called **algebraic equations.** For example,

$$5x^3 + 2x^2 - 3x + 5 = 0$$

$$x^4 - x^3 + x^2 - x - 1 = 0$$

$$\frac{x^2}{2} + \frac{3x}{5} - 3 = 0$$

If $f(x)$ also contains trigonometric, logarithmic, exponential etc. in addition to x, the equation $f(x) = 0$ is known as **transcendental equation.** For example

$$(3x^2 - 5)\, e^{-3x} + \log (1 - x) = 0$$

$$x + \sin x + 1/x\, e^{2x} = 0$$

$$\text{Cos}\, x + \log (1 - x) - 3e^x = 0$$

While solving equations of the form $f(x) = 0$, the following properties of algebraic equations must be remembered.

(*i*) Every algebraic equation of nth degree, has n and only n roots, *i.e.*, the number of the roots of the equation is equal to the degree of the equation.

(*ii*) Complex roots of the equation, if any, occur in pairs, *i.e.*, if $(a + ib)$ is a root of the equation $f(x) = 0$, there must be another root $(a - ib)$.

(*iiii*) If $x = a$ is the root of the equation $f(x) = 0$ then $(x - a)$ must be a factor of the equation $f(x) = 0$.

(*iv*) The number of positive roots of an algebraic equation $f(x) = 0$ with real coefficients does not exceed the number of changes in the signs of the terms of the function $f(x)$.

Let the given algebraic equation be

$$x^5 - 4x^3 - 3x^2 + 4x - 3 = 0$$

The signs of the coefficients are:

$$+ - - + -$$

As the total number of changes are 3, the equation can not have more than three positive roots.

(*v*) The number of negative roots of an algebraic equation can not have more negative roots than the number of changes of sign of the terms in the function $f(x)$

Let the given algebraic equation be

$$x^4 - 4x^3 - 3x^2 + 4x - 3 = 0$$

then the signs of the coefficients of the function $f(-x)$ be:

$$+ + - - -$$

As there is only one change in signs, the equation does not have more than one negative roots.

Ruel (*iv*) and (*v*) stated above for finding out the upper limit of positive and negative roots of $f(x) = 0$ are known as *Descartes rule of Signs.*

(*vi*) For locating the interval in which the roots of $f(x) = 0$ may lie, the following rules are applicable:

If $f(x)$ is a continuous function in the interval (a, b) and if $f(a)$ and $f(b)$ have opposite signs, the equation $f(x) = 0$ has at least one real root lying in the interval (a, b).

If $f(a) < f(b)$ numerically, a is taken as the first approximation of the roots otherwise b is taken as the first approximation.

Suppose $f(2) = -1.357$ and $f(3) = 4.283$.

The root of the equation will be between 2 and 3. As $f(2)$ is less than $f(3)$, the root will be nearer to 2. Hence, $x = 2$ is the first approximation of the root.

Consider the transcendental equation.

$$f(x) = e^{x-5} + x - 5 = 0$$

Here
$$f(2) = e^0 + 2 - 5 = -2 \text{ (negative)}$$
$$f(3) = e + 3 - 5 = 2.718 + 3 - 5 = 0.718 \text{ (positive)}$$

As $f(2)$ and $f(3)$ are of opposite signs, one root of the equation must be between 2 and 3.

Now, to decrease the interval, find the value of $f(2.5)$, *i.e.*,

$$f(2.5) = e^{0.5} + 2.5 - 5 = 1.649 + 25 - 5 = -0.851$$

As $f(2.5)$ and $f(3)$ are of opposite sings, the root of the given equation lies between 2.5 and 3.

5.11 SUCCESSIVE APPROXIMATION

The principle of successive approximation is used to solve the problems by assuming an initial approximation to the solution and then using a suitable formula. Attempts are made to improve the solution till the required accuracy is obtained. The principle of successive approximation is applied to a veriety of problems. Some of the methods using the principle of successive approximations are the following:

1. Newton-Rephason method to solve equations of the form $f(x) = 0$.
2. Gauss-Seidral method to solve linear simultaneous equations.
3. Pascal's method to solve differential equations.

These methods are explained in subsequent paragraphs.

1. Graphical Method: The approximate value of the roots of the equation $f(x) = 0$ may be obtained by the graphical method as discussed below.

(*i*) Draw the graph of $y = f(x)$.

(*ii*) The roots of the equation are the x-coordinate of the points at which y-coordinates are zero.

In case of polynomial $f(x)$ contains higher powers of x, the equation $f(x) = 0$ may be considered as $f_1(x) = f_2(x)$. Now plot two functions $y_1 = f_1(x)$ and $y_2 = f_2(x)$ on the same axes. The abscissa of the points of intersection of these curves are the real roots of the given equation $f(x) = 0$.

Example 5.14. *Solve the equation $x^3 + 3x - 7 = 0$, graphically.*

Solution: Let $f(x) = x^3 + 3x - 7$

Now $f(1) = 1 + 3 \times 1 - 7 = \text{(negative)}$

$f(2) = 2^3 + 3 \times 2 - 7 = \text{(positive)}$

i.e., the root of the equation lies between 1 and 2.

The equation can be written as

$$x^3 = 7 - 3x$$

keeping highest power of constants on the L.H.S.

To draw the graphs of $y = x^3$ and $y = 7 - 3x$,. from $x = 1$ to $x = 2$, we prepare the following tables:

Table 1 for function $y = x^3$

x	1	1.2	1.4	1.6	1.8	2
y	1	1.73	2.74	4.10	5.83	8

Table 2 for function $y = 7 - 3x$

x	1	1.4	1.6	2
y	4	2.8	2.2	1

After plotting these the graphs of the functions, we notice that they intersect at a point whose x coordinate is 1.4 (Fig. 5.1)

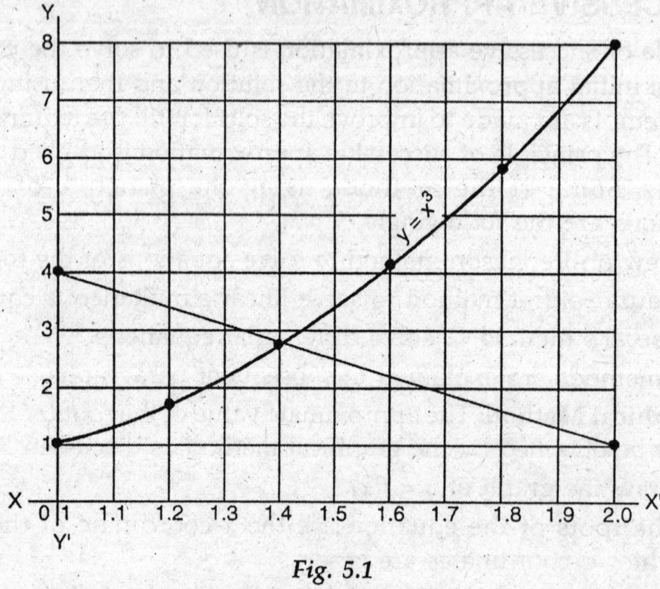

Fig. 5.1

Check. $x^3 + 3x - 7 = 1.4^3 + 3 \times 1.4 - 7$

$= 2.744 + 4.2 - 7$

$= 0.056 = 0.$ **O.K.**

Example 5.15. *Find graphically the positive root of the equation*

$$x^3 - 6x - 14 = 0$$

Solution: Let the given equation be represented as under:

$$f_{(x)} = x^3 - 6x - 13 = 0 \quad \text{...(}i\text{)}$$

$$f_{(2)} = 8 - 12 - 13 = -17, \quad -\text{ve}$$

$$f_{(3)} = 27 - 18 - 13 = -4, \quad -\text{ve}$$

$$f_{(4)} = 64 - 24 - 13 = 27, \quad +\text{ve}$$

From the above equations, we conclude that the root of eqn. (*i*) lies between 3 and 4 as $f_{(3)}$ and $f_{(4)}$ are opposite in sign.

Equation (*i*) may be written as under:

$$x^3 = 6x + 13$$

or

$$y = x^3 \text{ and } y = 6x + 13$$

Draw the graphs of equations $y = x^3$ and $y = 6x + 13$

Table 1. formation $y = x^3$

x	3	3.2	3.4	3.6	3.8	4.0
y	27	32.8	39.3	46.7	54.9	64

Table 2. formation $y = 6x + 13$

x	3	3.2	3.4	3.6	3.8	4
y	31	32.2	33.4	34.6	35.8	37

Taking (3, 0) as the origin of the cartisian coordinates, plot the graphs of $y = x^3$ and $y = 6x + 13$ as shown in Fig. 5.2.

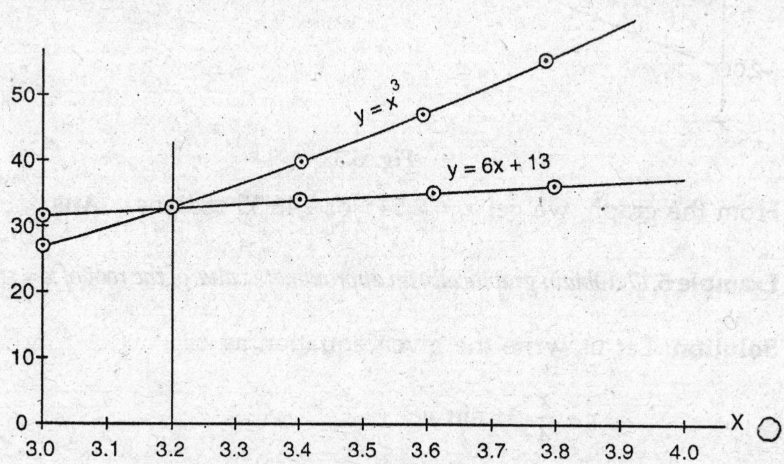

Fig. 5.2

The root of eqn (*i*) is the abscissa of the point of inter section of two graphs *i.e.*, 3.1. **Ans.**

Example 5.16. *Solve graphically the following equation:*

$$x - 2 = \sin x$$

Solution: The given equation $x - 2 = \sin x$ may be written as

$$y = x - z \text{ and } y = \sin x$$

Plot the graphs of the equations the find the abscissa of the point of intersection of the line $y = x - 2$ and the curve $y = \sin x$

function $y = x - 2$

x	0	1	2
y	− 2	− 1	0

function $y = \sin x$

x	0	$\pi/4$	$\pi/2$	$3\pi/4$	π
y	0	0.71	1.00	0.71	0

The graphs of the equations are shown in Fig. 5.3.

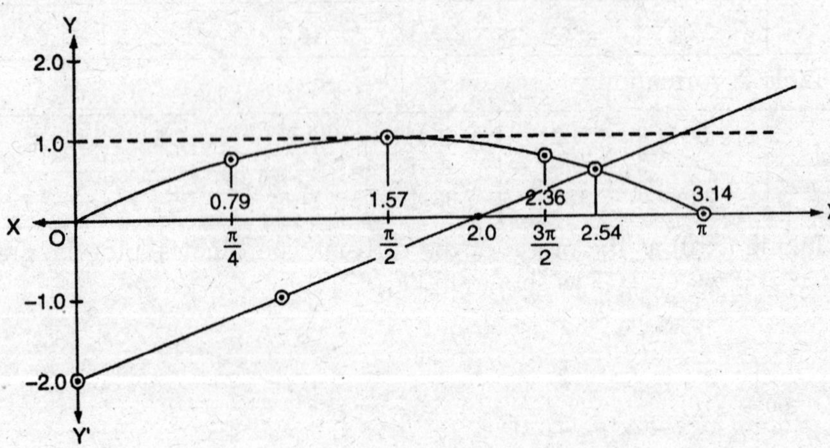

Fig. 5.3

From the graph, we get $x = 2.54$ or 46.35 radians. **Ans.**

Example 5.17. *Obtain graphically an approximate value of the root of* $x = \sin x + \frac{x}{4}$.

Solution: Let us write the given equation as

$$x - \frac{x}{4} = \sin x$$

The abscissa of the point of intersection of the curve $y = \sin x$ and the line $y = x - \frac{x}{4}$, will give a rough estimate of the root.

To draw the curve $y = \sin x$ and line $y = x - \frac{x}{4}$ we tabulate the values as under:

Table for the function $y = \sin x$

x(rad)	0	0.785	1.571	2.356	3.14
y	0	0.707	1	0.707	0

Table for the function $y = x - \frac{\pi}{4}$

x	0	3
y	1.57	4.57

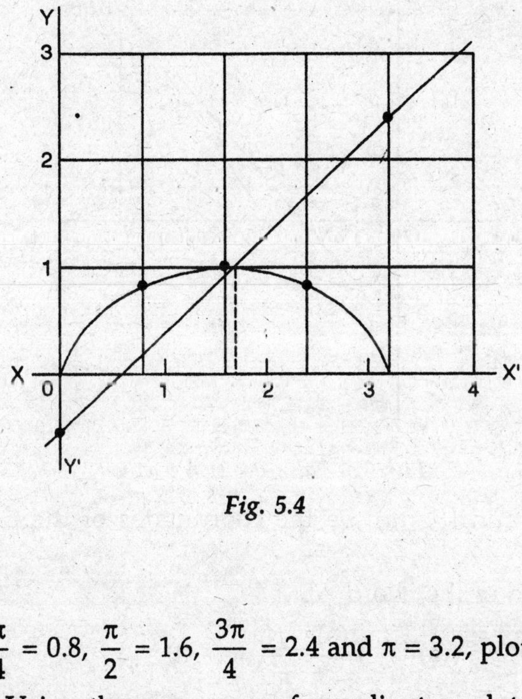

Fig. 5.4

By taking $\frac{\pi}{4} = 0.8$, $\frac{\pi}{2} = 1.6$, $\frac{3\pi}{4} = 2.4$ and $\pi = 3.2$, plot the graph of the curve $y = \sin x$. Using the same axes of coordinates, plot the graph of the straight line $y = x - \frac{\pi}{4}$.

From the point of intersection of the graphs, we get the value of x as 1.7 radians.

2. **False position (Regula falsi) method.** In this method, two numbers x_1 and x_2 are chosen as close as possible, and the point is determined between which the root lies. Since the root lies between x_1 and x_2, the graph of $y = f(x)$ must cross the x-axis between $x = x_1$ and $x = x_2$ and apparently y_1 and y_2 must have opposite signs.

The method of false position is based on the following principle.

As any small portion of a smooth curve may be assumed as a straight line, the change in $f(x)$ may be considered proportional to the change in x over that small portion.

Let the given equation be $f(x) = 0$.

Let $y = f(x)$ be represented by the curve AB.

Let the curve cut the X-axis at Q.

The real root of the curve is OQ. As we consider the portion of the curve AB as a straight line i.e. chord AB cuts the X-axis at P.

Hence, the approximate root of the curve $f(x) = 0$ is OP. (Fig. 5.5)

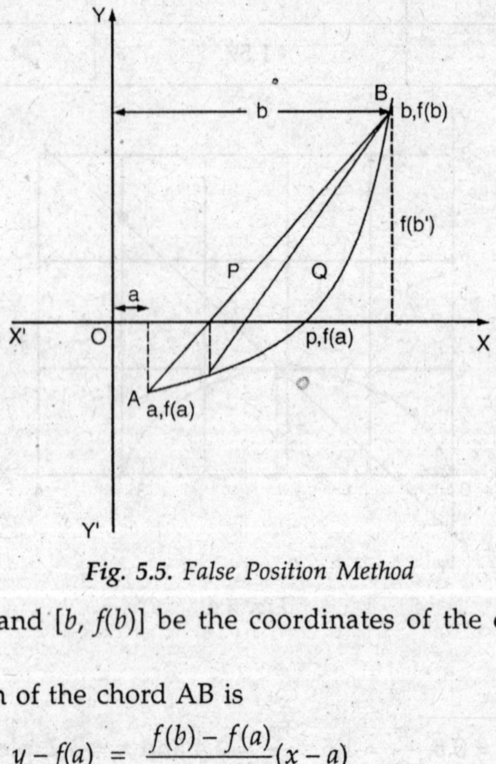

Fig. 5.5. False Position Method

Let $[a, f(a)]$ and $[b, f(b)]$ be the coordinates of the extremities of the chord AB.

The equation of the chord AB is

$$y - f(a) = \frac{f(b) - f(a)}{b - a}(x - a) \quad \ldots(i)$$

To obtain the location of P, put $y = 0$ in eqn. (i)

$$0 - f(a) = \frac{f(b) - f(a)}{b - a}(x - a)$$

or
$$(x - a) = \frac{-f(a) - [b - a]}{f(b) - f(a)}$$

or
$$x = a + \frac{f(a) - [a - b]}{f(b) - f(a)}$$

$$= \frac{af(b) - af(a) + af(a) - bf(a)}{f(b) - f(a)}$$

or
$$x = \frac{af(b) - bf(a)}{f(b) - f(a)}$$

i.e., The abscissa OP $= x = \dfrac{af(b) - bf(a)}{f(b) - f(a)}$.

The process is continued till we get the required value of the root.

Example 5.18. *Use Regula False method to find a real root of the equation $x \log_{10} x - 1.3 = 0$ correct to four places of decimal.*

Solution: Let $f(x) = x \log_{10} x - 1.3$

Now
$$f(2) = 2 \log_{10} 2 - 1.3$$
$$= 2 \times 0.30103 - 1.3 = -0.69794$$
$$f(3) = 3 \log_{10} 3 - 1.3$$
$$= 3 \times 0.47712 - 1.3 = 0.13136$$

As $f(2)$ and $f(3)$ are of opposite signs, there must be at least one root of the given equation which lies beween 2 and 3.

$$x_1 = \frac{af(b) - bf(a)}{f(b) - f(a)}$$

$$= \frac{2 \times 0.13136 - 3(-0.69794)}{0.13136 - (-0.69794)} = \frac{0.26272 + 2.09382}{0.13136 + 0.69794}$$

$$= \frac{2.35654}{0.8293} = +2.8416$$

Now
$$f(2.8416) = 2.8416 \times \log_{10} 2.8416 - 1.3$$
$$= 2.8416 \times 0.453563 - 1.3$$
$$= -0.01116$$

As $f(2.8416)$ and $f(3)$ are of opposite signs, the root lies in the interval of 2.8416 and 3.

The second approximate value

$$x_2 = \frac{2.8416 \times 0.13136 - 3(-0.01116)}{0.13136 + 0.01116}$$

$$= \frac{0.3732725 + 0.03348}{0.13136 + 0.01116} + \frac{0.4067525}{0.142252} = 2.8540$$

Now
$$f(2.8540) = 2.854 \log_{10} (2.854) - 1.3$$
$$= 1.299866 - 1.300000 = -0.000134$$

As $f(2.854)$ and $f(3)$ are of opposite signs, the root lies in the interval of 2.854 and 3.

The third approxmate value

$$x_3 = \frac{2.854 \times 0.13136 - 3(-0.000134)}{0.13136 + 0.000134}$$

$$= \frac{0.3749014 + 0.000402}{0.131360 + 0.000134} = \frac{0.3753034}{0.131494} = 2.8541$$

Now $f(2.8541) = 2.8541 \log_{10} (2.8541) - 1.3$

$= 1.299955 - 1.3 = -0.000045$

As $f(2.8541)$ and $f(3)$ are of opposite signs, the root lies in the interval 2.8541 and 3.

The fourth approximate value

$$x_4 = \frac{2.8541 \times 0.13136 - 3(-0.000045)}{0.13136 + 0.000045} = \frac{0.3750495}{0.131405}$$

$$= 2.85415$$

Hence, 2.85415 is the required value of the root correct to 5 places of decimal.

This is also an iteration method for solving equations $f(x) = 0$ where f is assumed to have a continuous derivative f. The main principle of the method is to approximate the graph of f by suitable tangents.

Example 5.19. *Determine the approximate value of the root of the equation $x^3 + 2x - 1$, using the method of false position (regula falsi).*

Solution: $f(x) = x^3 + 2x - 1$

Let the root of the equation

$$f(0.5) = (0.5)^3 + 2(0.5) - 1 = 0.125$$

Let the root of the equation be (0.2)

$$f(0.2) = (0.2)^3 + 2(0.2) - 1 = -0.592$$

Since the values of root are of opposite sign, the root lies between (0.2) and (0.5)

Let $x_1 = 0.2$ and $x_2 = 0.5$

$$x = \frac{x_1 f(x_2) - x_2 f(x_1)}{f(x_2) - f(x_1)} = \frac{0.2(0.125) - 0.5(-0.592)}{0.125 + 0.592}$$

$$= \frac{0.2 \times 0.125 + 0.5 \times 0.592}{0.717}$$

$$= \frac{0.025 + 0.296}{0.717} = 0.4477$$

Now $f(0.4477) = -0.0149$ and $f(0.5) = 0.125$

Root lies between 0.4477 abnd 0.5000 ·

$$x_1 = 0.4477, x_2 = 0.50000$$

$$x = \frac{0.4477(0.5) - 0.5(-0.0149)}{0.5 + 0.149} = \frac{0.223 + 0.0075}{0.5149}$$

$$x = 0.4494 \quad \textbf{Ans.}$$

Check: $(0.4494)^3 + 2 \times 0.4494 - 1$

$$= 0.09076 + 0.8988 - 1$$

$$= 0.98956 - 1 = -0.01044 \qquad \textbf{O.K.}$$

Example 5.20. *Find the root of the equation $2x - \log_{10} x = 7.2$ which lies between 3.5 and 4, correct to five places of decimal using method of false position.*

Solution: The given equation is $2x - \log_{10} x = 7.2$

or $\quad 2x - \log_{10} x - 7.2 = 0$

$$f(x) = 2x - \log_{10} x - 7.2$$

No, the value of the function $f(4)$ is given by

$$f(4) = 8 - \log_{10} 4 - 7.2 = 0.8 - 0.60206 = +0.19794$$

$$f(3.5) = 7 - \log_{10} 3.5 - 7.2 = 0.2 - 0.54407 = -0.34407$$

Since the valuesof $f(4)$ and $f(3.5)$ are of opposite signs, the root of the equation lies between 3.5 and 4.

Applying the false position method-formula

$$x_1 = \frac{x_1 f(x_2) - x_2 f(x_1)}{f(x_2) - f(x_1)}$$

$$= \frac{3.5 f(4) - 4 f(3.5)}{f(4) - f(3.5)}$$

$$= \frac{3.5(0.19794) - 4(-0.34407)}{0.19794 - (-0.34407)}$$

$$x_1 = \frac{0.69279 + 1.37628}{0.54201} = 3.81740$$

Now $\quad f(3.81740) = 7.6348 - 0.58177 - 7.2 = -0.3829$

Again, applying False position method,

$$x_2 = \frac{3.81740 f(4) - 4 f(3.81740)}{f(4) - f(3.81740)}$$

$$= \frac{3.81740 \times 0.19794 - 4(-0.3829)}{0.19794 - f(-0.3829)}$$

$$x_2 = \frac{0.75562 + 1.5316}{0.58084} = \frac{2.28722}{0.58084} = 3.88718$$

$\therefore$ Approximate value of the root = 3.88718 **Ans.**

Check: $2 \times 3.8878 - \log_{10}(3.88718)$

$7.77436 - 0.58963 = 7.18473 = 7.2$ **O.K.**

Example 5.21. *Calculate the approximate value of the root of the equation $x^3 - 3x + 3 = 0$, which lies between – 2 and – 3, correct to three places of decimals, using the method of false position.*

Solution: $f(x) = x^3 - 3x + 3 = 0$

Given: $x_1 = -2 \;;\; x_2 = -3$

$$f(x_1) = f(-2) = (-2)^3 - 3(-2) + 3 = -8 + 6 + 3 = 1$$

$$f(x_2) = f(-3) = (-3)^3 - 3(-3) + 3 = -27 + 9 + 3 = -15$$

$$x = \frac{x_1 f(x_2) - x_2 f(x_1)}{f(x_2) - f(x_1)}$$

$$= \frac{-2(-15) - (-3)(1)}{(-15) - (1)} = \frac{30 + 3}{-16} = \frac{33}{-16} = -2.063$$

Now $f(-2.063) = (-2.063)^3 - 3(-2.063) + 3$

$$= -8.780 + 6.180 + 3$$

$$= +0.409$$

$$f(-3) = -15$$

and $f(-2.063) = +.409$

$\therefore$ Root lies between – 2.063 and – 3

$$x = \frac{(-2.063)(-15) - (-3)(+0.409)}{(-15) - (0.409)}$$

$$= \frac{30.945 + 1.227}{-15.409} = \frac{32.172}{-15.409} = -2.088$$

Now, $f(-2.088) = (-2.088)^3 - 3(-2.88) + 3$

or $f(-2.205) = -9.103 + 6.264 + 3 = 0.161$

and $f(-3) = -15$

Root lies between – 3 and – 2.088

$$x = \frac{(-2.088)(-15) - (-3)(0.161)}{-15 - 0.161}$$

$$= \frac{31.32 + 0.483}{-15.161} = \frac{31.803}{-15.161}$$

or $x = -2.098$ **Ans.**

Check: $(-2.098)^3 - (2.098) + 3$

$$= -9.234 + 6.244 + 3 = 0.06 \quad \text{O.K.}$$

Example 5.22. *Ascertain the approximate root of the equation*

$$x^4 + x^3 - 7x^2 - x + 5 = 0$$

which lies between 2 and 3 correct to three decimal places.

Solution:

$$x^4 + x^3 - 7x^2 - x + 5 = 0$$

$$f(2) = 16 + 8 - 28 - 2 + 5 = -1 \quad \text{...}(i)$$

$$f(3) = 81 + 27 - 63 - 3 + 5 = +47 \quad \text{...}(ii)$$

From eqns. (*i*) and (*ii*) we notice that $f(2) = -1$. Hence the root of the equation is nearer to 2.

Let the root be 2.1.

$$f(2.1) = (2.1)^4 + (2.1)^3 - 7(2.1)^2 - (2.1) + 5$$

$$= 19.4481 + 9.2610 - 30.8700 - 2.1000 + 5$$

$$= +0.7391$$

Apparently the root lies between 2 and 2.1

Applying the method of False position,

$$x_1 = \frac{af(b) - bf(a)}{f(b) - f(a)} \text{ we get}$$

$$x_1 = \frac{2(0.7391) - 2.1(-1)}{(0.7391) - (-1)} = \frac{1.4782 + 2.1}{1.7391} = \frac{3.5782}{1.7391}$$

$$= 2.0575$$

Now, $$f(2.0575) = (2.0575)^4 + (2.0575)^3 - 7(2.0575)^2 - (2.0575) + 5$$

$$= 17.9209 + 8.7100 - 29.6331 - 2.0575 + 5 = -0.0587$$

$$x_2 = \frac{2.0575(0.7391) - 2.1(-0.0597)}{0.7391 - (-0.0597)}$$

$$= \frac{1.5207 + 0.1254}{0.7988} = \frac{1.6461}{0.7888}$$

$$= 2.0607$$

$$f(2.0607) = (2.0607)^4 + (2.0607)^3 - 7(2.0607)^2 - (2.0607) + 5$$

$$= 18.0326 + 8.7507 - 29.7254 - 2.0607 + 5$$

$$= -0.0028$$

$$x_3 = \frac{2.0607(0.7391) - 2.1(0.0028)}{0.7391 - (-0.0028)}$$

$$= \frac{1.5231 + 0.0059}{0.7419} = \frac{1.5290}{0.7419}$$

$$x_3 = 2.0609$$

The required root of the given equation s 2.0609 **Ans.**

Check: $f(x) = (2.0609)^4 + (2.0609)^3 - 7(2.0607)^2 - 2.0607 + 5$

$18.0396 + 8.7533 - 29.7312 - 2.0609 + 5 = 0.0008$

= very nearly equal to zero **Ans.**

3. The Newton-Raphson method. (The Newton Raphson method of polynomial equation). This method is used to determine the real roots of the equation $f(x) = 0$ if its derivative can be found easily. The underlying idea of the method was originaly originated by Newton and the method now used is due to Raphson. That's why the method is known as the Newton-Raphson method.

Derivation of the formula

Let the given equation be $f(x) = 0$

the exact value of the root be x

an approximate value of the desired root be c.

the correction to be applied to c be h.

$$\therefore \quad x = c + h.$$

Substituting the value $x = (c + h)$ in $f(x) = 0$, we get

$$f(c + h) = 0$$

Expanding $f(c + h)$ by Taylor's thorem, we get

$$f(c + h) = f(c) + hf'(x) + \frac{h^2}{2} f''(c) + \ldots = 0$$

As h is relatively small, we may ignore h^2 and other higher powers of h. The expanded form of the equation may be written as

$$f(c) + hf'(c) = 0$$

or
$$h_1 = \frac{-f(c)}{f'(c)}$$

The improved value of the root

$$c_1 = c + h_1 = c - \frac{f(c)}{f'(c)}$$

$$c_2 = c_1 + h_2 = c_1 - \frac{f(c_1)}{f'(c)}$$

$$c_3 = c_2 + h_3 = c_2 - \frac{f(c_2)}{f'(c_2)}$$

....

....

....

$$c_n = c_{n-1} + h_n = c_{n-1} - \frac{f(C_{n-1})}{f'(C_{n-1})}$$

From the fundamental formula of the Newton-Raphson, $h_1 = \frac{f(c)}{f'(c)}$, it is clear that if $f'(x)$ is larger, the necessary correction to be applied to get the correct value of the root is smaller. In other words, it is possible to find the value of the root rapidly without spending much time if the graph is nearly vertical where it crosses the x-axis. The Newton-Raphson method is not suitable if the graph of $f(x) = 0$ is nearly horizontal where it crosses the x-axis.

Geometric significance of the formula

Let the graph of $y = f(x)$ be represented as shown in Fig. 5.6.

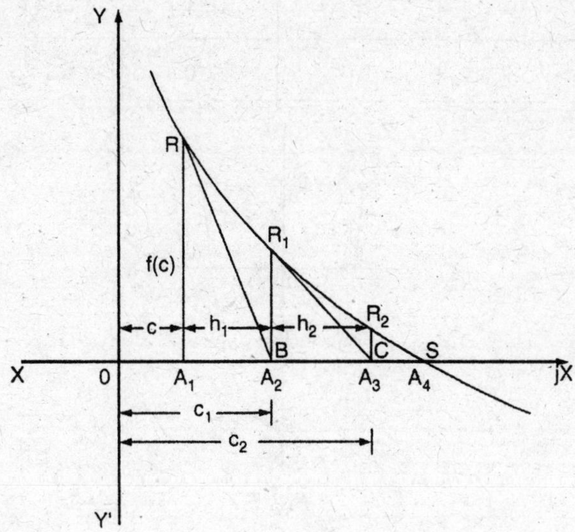

Fig. 5.6

Draw a tangent from any point R whose abscissa = c.

Let this tangent cut the X-axis at B.

Draw another tangent from R_1 whose abscissa is OB = c_1.

Let this tangent cut the X-axis at some point C between B and S.

We may now draw a tangent from R_2 whose abscissa is OC = c_2.

Let this tangent cut the X-axis at some point D between C and S. It is thus clear that if the curvature of the graph does not change sign between R and S, the points $A_1, A_2, A_3, A_4, \ldots$ approach the point S as a limit.

But, OS is the real root of the equation, whose graph is plotted in Fig. 3.4.

Hence, $OA_1, OA_2, OA_3, OA_4, \ldots OA_n$ be the successive approximations to the desired root.

Proof. Let $A_1A_2 = h_1$; $A_2A_3 = h_2, \ldots$

Let $\qquad A_1R = f(c)$

$$\text{Slope of curve at } R = \tan OA_2R = -\frac{f(x)}{h_1}$$

$$\therefore \qquad f'(c) = -\frac{f(c)}{h_1}, \quad \text{or} \quad h_1 = -\frac{f(c)}{f'(c)}$$

Example 5.23. *By the Newton_Raphson method, solve the equation*

$$3x^2 - \cos x - 1 = 0$$

Solution: $f(x) = 3x^2 - \cos x - 1$

$\therefore$ $f'(x) = 6x + \sin x$

For getting the rough value of the root, plot the graph of the function as shown below

x	0	1	– 1	– 0.7	2.2
y	– 2	1.0	1.0	– 1.75	4.75

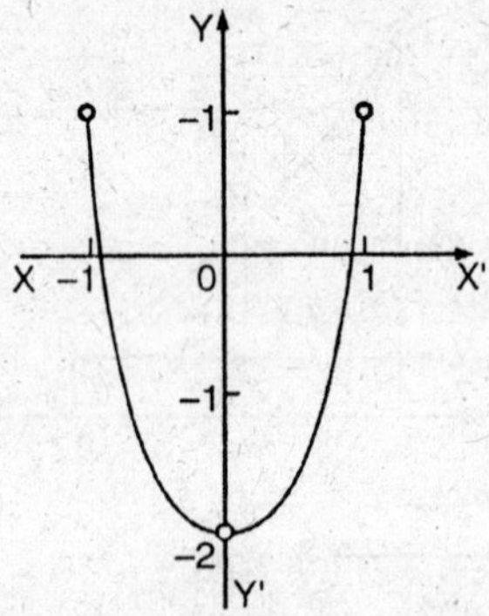

Fig. 5.7

Since the term x^2 is positive for all real values of x, it is evident that the equation will be satisfied only by a negative value of x.

Let us take the approximate negative value of the root from the graph as

$$x_0 = -0.65$$

$$\therefore \quad x_1 = x_0 - \frac{3x^2 - \cos x - 1}{6x + \sin x}$$

$$= -0.65 - \frac{3(-0.65)^2 - \cos(0.65) - 1}{6(-0.65) + \sin(-0.65)}$$

$$= -0.65 - \frac{1.2675 - 0.9999 - 1}{-3.9 - 0.6052}$$

$$= 0.65 - \frac{-0.7324}{-4.5052} = -0.65 - 0.187$$

$$\therefore \quad x_1 = -0.837$$

$$x_2 = x_1 - \frac{3(-0.837)^2 - \cos(-0.0837) - 1}{6(-0.837) + \sin(-0.837)}$$

$$= -0.837 - \frac{0.1018}{-5.0366}$$

$$= -0.837 + 0.020$$

or $$x_2 = -0.817$$

This value of the root is correct to its last figure. **Ans.**

Example 5.24. *Solve the following simultaneous equationsusing Newton-Raphson method:*

$$x^2 + y = 28$$
$$2x - y^2 = 1.$$

Solution: Plot the graphs of the equations on the same set of axes to get the two set of values (5, 3) and (5.4, 3.2).

We shall improve the second set of the values.

Let $$f_1(x_1, y) = x^2 + y - 28$$
$$f_2(x, y) = 2x - y^2 - 1$$

$\therefore$ $$\frac{\partial f_1}{\partial x} = 2x$$
$$\frac{\partial f_1}{\partial y} = 1$$
$$\frac{\partial f_2}{\partial x} = 2$$
$$\frac{\partial f_2}{\partial y} = -2y$$

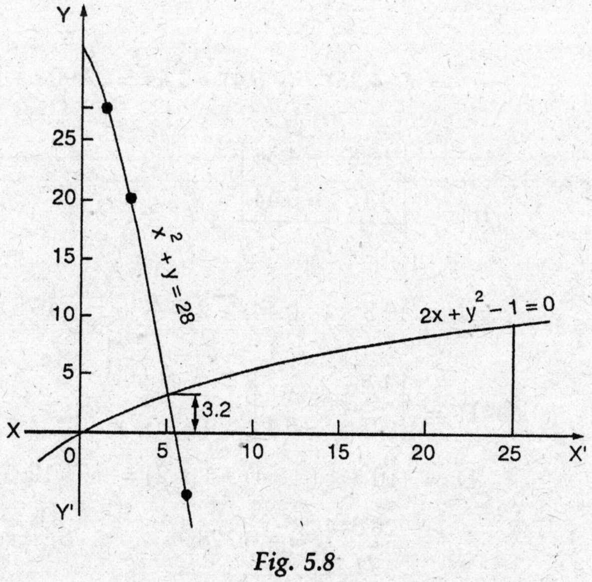

Fig. 5.8

Now we know that $x_0 = 5.4$ and $y_0 = 3.2$

$$\therefore \quad f_1(x_0, y_0) = (5.4)^2 + 3.2 - 28 = 4.36$$

x	1	3	6	8
y	27	19	– 8	– 36

x	0	5	13	25
y	1	3	4	7

$$f_2(x_0, y_0) = 2 \times 5.4 - 3.2^2 - 1 = -0.44$$

$$\therefore \quad \frac{\partial f_1}{\partial x} = 2 \times 5.4 = 10.8$$

$$\frac{\partial f_1}{\partial y} = 1$$

$$\frac{\partial f_2}{\partial x} = 2$$

$$\frac{\partial f_2}{\partial y} = -2 \times 3.2 = -6.4$$

Substituting the values in eqns. (9), (10) and (11), we get

$$h_1^{(1)} = \frac{\begin{vmatrix} -4.36 & 1 \\ -0.44 & -6.4 \end{vmatrix}}{D}$$

$$= (-4.36) \times (-6.4) + 0.44 = 27.904 + 0.44 = \frac{28.344}{D}$$

$$h_2^{(1)} = \frac{\begin{vmatrix} 10.8 & -4.36 \\ 2 & -0.44 \end{vmatrix}}{D}$$

$$= 10.8 \times (-0.44) + 2 \times 4.36 = -4.75 + 8.72 = \frac{3.97}{D}$$

$$D = \begin{vmatrix} 10.8 & 1 \\ 2 & -6.4 \end{vmatrix}$$

$$D = (10.8 \times (-6.4) - 1 \times 2) = -71.12$$

$$h_1 = \frac{28.344}{-71.12} = -0.3985$$

$$h_2 = \frac{3.97}{-71.12} = -0.0558$$

Hence $x^{(1)} = 5.4 - 0.4 = 5$

$y^{(1)} = 3.2 - 0.06 =$ **3.14** **Ans.**

5.12 THE NEWTON-RAPHSON METHOD OF SIMULTANEOUS EQUATIONS

Let the simultaneous equations of two unknowns be:

$$f_1(x, y) = 0 \qquad ...(1)$$

$$f_2(x, y) = 0 \qquad ...(2)$$

Let the approximate values of a pair of roots be x_0, y_0 and the respective corrections be h_1 and h_2.

Substituting the values in equations (1) and (2), we get

$$f_1(x_0 + h_1, y_0 + h_2) = 0 \qquad ...(3)$$

$$f_2(x_0 + h_1, y_0 + h_2) = 0 \qquad ...(4)$$

Expanding he eqns. (3) and (4) by Taylor's theorem for the function of two variables, we get

$f_1(x_0 + h_1, y_0 + h_2) = f_1(x_0, y_0) + h_1\left(\frac{\partial f_1}{\partial x}\right)_0 + h_2\left(\frac{\partial f_1}{\partial y}\right)_0$ + terms containing higher powers of h_1 and h_2 ...(5)

and, $f_2(x_0 + h_1, y_0 + h_2) = f_2(x_0, y_0) + h_1\left(\frac{\partial f_2}{\partial x}\right)_0 + h_2\left(\frac{\partial f_2}{\partial y}\right)_0$ + terms containing higher powers of h_1 and h_2. ...(6)

Since h_1 and h_2 are relatively small, their squares and products and higher powers may be ignored.

The eqns. (5) and (6) thus reduce to:

$$f_1(x_0, y_0) + h_1\left(\frac{\partial f_1}{\partial x}\right)_0 + h_2\left(\frac{\partial f_1}{\partial x}\right)_0 = 0 \qquad ...(7)$$

and $$f_2(x_0, y_0) + h_1\left(\frac{\partial f_2}{\partial x}\right)_0 + h_2\left(\frac{\partial f_2}{\partial x}\right)_0 = 0 \qquad ...(8)$$

Solving equations (7) and (8) by using determinants, we get

$$h_1 = \frac{\begin{vmatrix} -f_1(x_0, y_0) & \left(\frac{\partial f_1}{\partial y}\right)_0 \\ -f_2(x_0, y_0) & \left(\frac{\partial f_2}{\partial y}\right)_0 \end{vmatrix}}{D} \qquad ...(9)$$

$$h_2 = \frac{\begin{vmatrix} \left(\frac{\partial f_1}{\partial x}\right)_0 & -f_1(x_0, y_0) \\ \left(\frac{\partial f_2}{\partial x}\right)_0 & -f_2(x_0, y_0) \end{vmatrix}}{D} \qquad ...(10)$$

where
$$D = \begin{vmatrix} \left(\frac{\partial f_1}{\partial x}\right)_0 & \left(\frac{\partial f_1}{\partial y}\right)_0 \\ \left(\frac{\partial f_2}{\partial x}\right)_0 & \left(\frac{\partial f_1}{\partial y}\right)_0 \end{vmatrix} \qquad ...(11)$$

Further corrections may be found out by repeated application of these formulae using improved values of h_1 and h_2 in each case.

5.13 SIMULTANEOUS ALGEBRAIC EQUATIONS

The simultaneous linear equations (more than 3) can be solved by two methods commonly known as the direct method and iterative method.

(1) **The Gauss Elimination method.** This method is a direct method of solving simulataneous equations. Consider the following three simultaneous equations with three unknowns.

$$a_{11}x_1 + a_{12}x_2 + a_{13}x_3 = a_{14} \qquad ...(1)$$
$$a_{21}x_1 + a_{22}x_2 + a_{23}x_3 = a_{24} \qquad ...(2)$$
$$a_{31}x_1 + a_{32}x_2 + a_{33}x_3 = a_{34} \qquad ...(3)$$

1st Step. To eliminate the first term from equation (2), divide eqn. (1) by a_{11} and multiply by a_{21}, *i.e.*,

$$a_{21} \times \frac{a_{11}}{a_{11}} x_1 + a_{21} \times a_{21} \times \frac{a_{12}}{a_{11}} x_2 + a_{21} \times a_{21} \times \frac{a_{13}}{a_{11}} x_3 + a_{21} \times \frac{a_{14}}{a_{11}} \qquad ...(1a)$$

$$a_{21}x_1 + a_{22}x_2 + a_{23}x_3 = a_{24} \qquad ...(2a)$$
$$a_{31}x_1 + a_{32}x_2 + a_{33}x_3 = a_{34} \qquad ...(3a)$$

Substracting eqn. (1a) from eqn. (2), we get

$$\left(a_{21} - \frac{a_{21} \times a_{11}}{a_{11}}\right)x_1 + \left(a_{22} - \frac{a_{21} \times a_{12}}{a_{11}}\right)x_2 + \left(a_{23} - \frac{a_{21} \times a_{13}}{a_{11}}\right)x_3 = a_{24} - \frac{a_{21} \times a_{14}}{a_{11}}$$

or $(a_{21} - k_2a_{11})x_1 + (a_{22} - k_2a_{12})x_2 + (a_{23} - k_2a_{13})x_3 + (a_{24} - k_2a_{14})$ ...(2a)

Similarly multiply eqn. (1) by $\frac{a_{21}}{a_{11}} = k_3$ and subtract it from eqn. (3) *i.e.*,

$$(a_{31} - k_3a_{11})x_1 + (a_{32} - k_3a_{12})x_2 + (a_{33} - k_3a_{13})x_3 + (a_{34} - k_3a_{14}) \quad ...(3a)$$

It may be clearly understood that $(a_{21} - k_2a_{11})$ and $(a_{31} - k_3a_{11})$ are both zero.

Now, by similarl process, eliminate x_2 from eqn. (3).

This method will be understood by following the solution of Example 5.?.

Example 5.25. *Solve the following system of simultaneous equations by the Gauss-elimination method or Gauss elimination method showing calculations for four successive iterations:*

$$4x - y + z = 4$$
$$x + 6y + 2z = 9$$
$$-x - 2y + 5z = 2.$$

Solution: Given equations

$$4x - y + z = 4 \quad ...(1)$$
$$x + 6y + 2z = 9 \quad ...(2)$$
$$-x - 2y + 5z = 2 \quad ..(3)$$

Dividing eqn. (1) by 4 and subtracting the resultant equation from eqn. (2), we get eqn. (2*a*). Adding eqns. (2) and (3), we get eqn. (3*a*)

$$4x - y + z = 4 \quad ...(1a)$$
$$0 + \frac{25y}{4} + \frac{7}{4}z = 8 \quad ...(2a)$$
$$0 + 4y + 7z = 11 \quad ...(3a)$$

Multiplying eqn. (3*a*) by $\frac{25}{16}$ and subtracting (2*a*) from the resultant, we get

$$4x - y + z = 4 \quad ...(1b)$$
$$0 + \frac{25y}{4} + \frac{7}{4}z = 8 \quad ...(2b)$$
$$0 + 0 + \frac{147z}{16} = \frac{147}{16} \quad ...(3b)$$

$$\therefore \quad \frac{147z}{16} = \frac{147}{16}$$

or $$z = 1$$

Substituting $z = 1$ in eqn. (2*b*), we get

$$\frac{25}{4}y = 8 - \frac{7}{4} \times 1 = \frac{25}{4}$$

$$\therefore \quad y = 1$$

Substituting the value of $z = 1$ and $y = 1$ in eqn. (1*b*), we get

$$4x - 1 + 1 = 4$$
$$4x = 4$$
$$x = 1$$

$$\left.\begin{array}{l} x=1 \\ y=1 \\ z=1 \end{array}\right\} \textbf{Ans.}$$

(2) **Gauss-Jordan Method.** In this method, the co-efficients of the variables both above and below the pivot in a given column, are eliminated. By this method, the equations are diagonalized and hence the back substitution step is eliminated. This is also a direct method of solving the simultaneous equations.

To understand the procedure to be adopted, refer to the following solved example 5.?.

Example 5.26. *Solve the following simultaneous equations by Gauss-Jordan method:*

$$\begin{aligned} 4x_1 + 2x_2 - 3x_2 &= 2 \\ 3x_1 + x_2 - 2x_3 &= 1 \\ 2x_1 - 2x_2 + x_3 &= 2. \end{aligned}$$

Solution: The augmented matrix is

$$\left[\begin{array}{ccc|c} 4 & 2 & -3 & 2 \\ 3 & 1 & -2 & 1 \\ 2 & -2 & 1 & 2 \end{array}\right]$$

Normalize the first row, second row and third row by dividing by 4, 3 and 2 respectively so that the matrix becomes

$$\left[\begin{array}{ccc|c} 1 & \frac{1}{2} & -\frac{3}{4} & \frac{1}{2} \\ 1 & \frac{1}{3} & -\frac{2}{3} & \frac{1}{3} \\ 1 & -1 & \frac{1}{2} & 1 \end{array}\right]$$

To make the remaining elements of the first column as zero, subtract the first row from the second row and also subtract the first row from the third row *i.e.,*

$$\left[\begin{array}{ccc|c} 1 & \frac{1}{2} & -\frac{3}{4} & \frac{1}{2} \\ 0 & -\frac{1}{6} & +\frac{1}{12} & -\frac{1}{6} \\ 0 & -\frac{3}{2} & \frac{5}{4} & \frac{1}{2} \end{array}\right]$$

Multiply the second row by – 6 and the third row by $\frac{-2}{3}$, we get

$$\left[\begin{array}{ccc|c} 1 & \frac{1}{2} & -\frac{3}{4} & \frac{1}{2} \\ 0 & 1 & -\frac{1}{2} & 1 \\ 0 & 1 & -\frac{5}{4} & -\frac{1}{3} \end{array}\right]$$

Subtracting the second row from the third row, we get

$$\left[\begin{array}{ccc|c} 1 & \frac{1}{2} & -\frac{3}{4} & \frac{1}{2} \\ 0 & 1 & -\frac{1}{2} & 1 \\ 0 & 0 & -\frac{1}{3} & -\frac{4}{3} \end{array}\right]$$

Multiplying the third row by – 3, we get

$$\left[\begin{array}{ccc|c} 1 & \frac{1}{2} & -\frac{3}{4} & \frac{1}{2} \\ 0 & 1 & -\frac{1}{2} & 1 \\ 0 & 0 & 1 & 4 \end{array}\right]$$

Multiplying the second row by 1/2 and subtracting it from the first row, we get

$$\left[\begin{array}{cccc} 1 & 0 & -\frac{1}{2} & 0 \\ 0 & 1 & -\frac{1}{2} & 1 \\ 0 & 0 & 1 & 4 \end{array}\right]$$

Multiplying the third row by 1/2 and adding to the first row and also to the second row, we get

$$\left[\begin{array}{ccc|c} 1 & 0 & 0 & 2 \\ 0 & 1 & 0 & 3 \\ 0 & 0 & 1 & 4 \end{array}\right]$$

∴ The values of x_1, x_2 and x_3 are 2, 3 and 4. **Ans.**

Example 5.27. *Solve the following systems of simultaneous equations by Gauss-Jordan method:*

$$\begin{aligned} 4x - y + z &= 4 \\ x + 6y + 2z &= 9 \\ -x - 2y + 5z &= 2. \end{aligned}$$

Solution: The augmented matrix is

$$\left[\begin{array}{rrr|r} 4 & -1 & 1 & 4 \\ 1 & 6 & 2 & 9 \\ -1 & -2 & 5 & 2 \end{array}\right]$$

Normalise the first row by dividing by 4

$$\left[\begin{array}{rrr|r} 1 & -\frac{1}{4} & \frac{1}{4} & 1 \\ 1 & 6 & 2 & 9 \\ -1 & -2 & 5 & 2 \end{array}\right]$$

Subtract the first row from the second row and add the first row to the third row, we get

$$\left[\begin{array}{ccc|c} 1 & -\frac{1}{4} & \frac{1}{4} & 1 \\ 0 & \frac{25}{4} & \frac{7}{4} & 8 \\ 0 & -\frac{9}{4} & \frac{21}{4} & 3 \end{array}\right]$$

Multiplying second row by 4/25 and the third row by 4/9, we get

$$\left[\begin{array}{ccc|c} 1 & -\frac{1}{4} & \frac{1}{4} & 1 \\ 0 & 1 & \frac{7}{25} & \frac{32}{25} \\ 0 & -1 & \frac{7}{3} & \frac{4}{3} \end{array}\right]$$

Adding the second row to the third row, we get

$$\left[\begin{array}{ccc|c} 1 & -\frac{1}{4} & \frac{1}{4} & 1 \\ 0 & 1 & \frac{7}{25} & \frac{32}{25} \\ 0 & -1 & \frac{196}{75} & \frac{196}{75} \end{array}\right]$$

Normalizing the third row, we get

$$\left[\begin{array}{ccc|c} 1 & -\frac{1}{4} & \frac{1}{4} & 1 \\ 0 & 1 & \frac{7}{25} & \frac{32}{25} \\ 0 & 0 & 1 & 1 \end{array}\right]$$

Multiplying the second row by 1/4 and adding to the first row

$$\left[\begin{array}{ccc|c} 1 & 0 & \frac{32}{100} & \frac{33}{25} \\ 0 & 1 & \frac{7}{25} & \frac{32}{25} \\ 0 & 0 & 1 & 1 \end{array}\right]$$

Multiplying the third row by 32/100 and subtracting it from the first row

$$\left[\begin{array}{ccc|c} 1 & 0 & 0 & 1 \\ 0 & 1 & \frac{7}{25} & \frac{32}{25} \\ 0 & 0 & 1 & 1 \end{array}\right]$$

Multiplying the third row by 7/25 and subtracting it from the second row

$$\left[\begin{array}{ccc|c} 1 & 0 & 0 & 1 \\ 0 & 1 & 0 & 1 \\ 0 & 0 & 1 & 1 \end{array}\right]$$

$$\therefore \qquad \left.\begin{array}{l} x_1 = 1 \\ x_2 = 1 \\ x_3 = 1 \end{array}\right\} \quad \textbf{Ans.}$$

3. **Gauss-Seidel iteration method.** Consider the following simultaneous equations:

$$a_{11}x_1 + a_{12}x_2 + a_{13}x_3 = d_1 \qquad \text{...(1)}$$

$$a_{21}x_1 + a_{22}x_2 + a_{23}x_3 = d_2 \qquad \text{...(2)}$$

$$a_{31}x_1 + a_{32}x_2 + a_{33}x_3 = d_3 \qquad \text{...(3)}$$

....

....

....

$$a_{n1}x_1 + a_{n2}x_2 + a_{n3}x_3 = d_n \qquad \text{...}(n)$$

In this method, the values of the unknowns in equation (1) are assumed as zero except th value of the first unknown and the velue of the first unknown is computed.

Next, substituting the value of the first unknown as calculated above and the values of 3rd, 4th, 5th etc., as zero, in the second equation, compute the value of second unknown (x_2). Proceed in this manner till the values of all the unknowns are computed. This is end of first *iteration.*

Now, substituting the first approximations of the values of second, 3rd, 4th, etc., in the first equation, compute the value of the first unknown and similarly, compute the values of other unknowns using the first approximations.

Repeat the required number of iterations till the successive values of x_1, x_2, x_3 ... x_n are close enough for our requirement.

The method is explained by solved example 5.28.

Example 5.28. *Solve the following system of simultaneous equations by the Gauss-Seidel iterative method or Gauss elimination method showing calculations for four successive iterative*

$$4x - y + z = 4$$
$$x + 6y + 2z = 9$$
$$-x - 2y = 5z = 2$$

Assume $x = y = z = 0$ as the initial guess.

Solution:

$$4x - y + z = 4 \quad ...(1)$$
$$x + 6y + 2z = 9 \quad ...(2)$$
$$-x - 2y = 5z = 2 \quad ...(3)$$

Substituting $y = 0$ and $z = 0$ in eqn. (1), we get

$$4x = 4$$
$$\therefore \quad x = 1$$

Substituting $x = 1$, $z = 0$ in eqn. (2), we get

$$1 + 6y + 0 = 9$$

or
$$y = \frac{8}{6} = \frac{4}{3}$$

Substituting $x = 1$, $y = 4/3$ in eqn. (3), we get

$$-1 - \frac{8}{3} + 5z = 2$$
$$5z = 2 + \frac{11}{3} = \frac{17}{3}$$
$$z = \frac{17}{3} \times \frac{1}{5} = \frac{17}{15}$$

This is the end of first iteration.

Substituting the value of $y = \frac{4}{3}$ and $z = \frac{17}{15}$ in eqn. (1)

$$4x - \frac{4}{3} + \frac{17}{15} = 4$$

$$4x = 4 + \frac{4}{3} - \frac{17}{15} = \frac{60+20-17}{15} = \frac{63}{15}$$

$$\therefore \quad x = \frac{63}{15} \times \frac{1}{4} = \frac{63}{60}$$

Substituting $x = y = \frac{63}{60}$ and $z = \frac{17}{15}$ in eqn. (2), we get

$$\frac{63}{60} + 6y + 2 \times \frac{17}{15} = 9$$

$$6y = 9 - \frac{63}{60} - \frac{34}{15} = \frac{540-63-136}{60}$$

$$6y = \frac{341}{60}$$

or $$y = \frac{341}{60} \times \frac{1}{6} = \frac{341}{360}$$

Substituting $x = \frac{63}{60}$ and $y = \frac{341}{360}$ is eqn. (3), we get

$$-\frac{63}{60} - 2 \times \frac{341}{360} + 5z = 2$$

$$\therefore \quad 5z = 2 + \frac{63}{60} + \frac{341}{180} = \frac{360+189+341}{180} = \frac{890}{180}$$

$$z = \frac{89}{18} \times \frac{1}{5} = \frac{89}{90}$$

This is end of second iteration.

Substituting $y = \frac{341}{360}$ and $z = \frac{89}{90}$ in eqn. (1) we get

$$4x - \frac{341}{360} + \frac{89}{90} = 4$$

$$\therefore \quad 4x = 4 + \frac{341}{360} - \frac{89}{90} = \frac{1440+341-356}{360} = \frac{1425}{360}$$

$$\therefore \qquad x = \frac{1425}{360} \times \frac{1}{4} = \frac{1425}{1440}$$

Substituting $x = \dfrac{1425}{1440}$ and $z = \dfrac{89}{90}$ in eqn. (2), we get

$$\frac{1425}{1440} + 6y + 2 \times \frac{89}{90} = 9$$

$$6y = 9 - \frac{1425}{1440} - \frac{178}{90} = \frac{12960 - 1425 - 2848}{1440}$$

$$\therefore \qquad y = \frac{8687}{1440} \times \frac{1}{6} = \frac{8687}{8640}$$

Substituting $x = \dfrac{1425}{1440}$ and $y = \dfrac{8687}{8640}$ in eqn. (3), we get

$$\frac{1425}{1440} - 2 \times \frac{8687}{8640} + 5z = 2$$

$$5z = 2 + \frac{1425}{1440} + \frac{17374}{8640} = \frac{17280\_8550 + 17374}{8640}$$

$$= \frac{43204}{8640}$$

$$\therefore \qquad z = \frac{43204}{8640} \times \frac{1}{5} = \frac{43204}{43200} = 1$$

This is the end of third iteration.

The fourth iteration may be carried out in a similar way to get the improved values;

$$\left.\begin{aligned} x &= 1 \\ y &= 1 \\ z &= 1 \end{aligned}\right\} \quad \textbf{Ans.}$$

Example 5.29. *Solve the following equations by Gauss-siedel integration method, using only three iterations.*

$$\begin{aligned} 10x_1 + 2x_2 + x_3 &= 9 \\ 2x_1 + 20x_2 - 2x_3 &= -44 \\ -2x_1 + 3x_2 + 10x_3 &= 22 \end{aligned}$$

Solution: $$x_1 = \frac{1}{10}(9 - 2x_2 - x_3) \qquad ...(a)$$

$$x_2 = \frac{1}{20}(-44 - 2x_1 + 2x_3) \quad ...(b)$$

$$x_3 = \frac{1}{10}(22 + 2x_1 - 3x_3) \quad ...(c)$$

Let $x_1 = 0, x_2 = 0, x_3 = 0$

Substituting the values in eqns. (*a*), (*b*), and (*c*), we get

$$x_1^{(1)} = \frac{1}{10}(9 + 0 + 0) = 9/10 = 0.9$$

$$x_2^{(1)} = \frac{1}{20}(-44 - 2 \times 0.9 + 0) = \frac{-45.8}{20} = -2.29$$

$$x_3^{(1)} = \frac{1}{10}[22 + 2 \times .9 - 3 \times (-2.29)] = \frac{30.67}{10} = 3.067$$

This is the end of the first iteration.

$$x_1^{(2)} = \frac{1}{10}(9 - 2x_2 - x_3) = \frac{1}{10}(9 + 2 \times 2.29 - 3.067) = 1.051$$

$$x_2^{(2)} = \frac{1}{20}(-44 - 2x_1 + 2x_3)$$

$$= \frac{1}{20}(-44 - 2 \times 1.051 + 2 \times 3.067) = -1.998$$

$$x_3^{(2)} = \frac{1}{10}(22 + 2 \times 1.051 + 3 \times 1.988) = 3.010$$

This is the end of second iteration

$$x_1^{(3)} = \frac{1}{10}(9 + 2 \times 1.998 - 3.0101) = 0.999 = 1.00$$

$$x_2^{(3)} = \frac{1}{20}(-44 - 2 \times 0.999 + 2 \times 3.01) = -1.999 = -2.00$$

$$x_3^{(3)} = \frac{1}{10}[22 + 2 \times 0.999 - 3 \times (-1.999)]$$

$$= \frac{1}{10}(22 + 1.998 + 5.9\ldots) = 2.9995 = 3.00$$

This is the end of third iteration

$$\begin{bmatrix} x_1 = 1 \\ x_2 = -2 \\ x_3 = 3 \end{bmatrix} \quad \textbf{Ans.}$$

Example 5.30. *Solve the following set of simultaneous equations using the Gaussian elmination method:*

$$3x - y + 5z = 16$$
$$x - 3y + 2z = 1$$
$$4x + y + z = 9$$

Solution:

$$3x - y + 5z = 16$$
$$x - 3y + 2z = 1$$
$$4x + y + z = 9$$

The augmented coefficient matrix is:

$$\begin{bmatrix} 3 & -1 & 5 & \vdots & 16 \\ 1 & -3 & 2 & \vdots & 1 \\ 4 & 1 & 1 & \vdots & 9 \end{bmatrix}$$

Replacing $R_2 \to \frac{1}{3}(3R_2 - R_1)$

$$\begin{bmatrix} 3 & -1 & 5 & \vdots & 16 \\ 0 & -2\frac{2}{3} & \frac{1}{3} & \vdots & -\frac{13}{3} \\ 4 & 1 & 1 & \vdots & 9 \end{bmatrix}$$

Replacing $R_3 \to \frac{3}{4}R_3 - R_1$ we get

$$\begin{bmatrix} 3 & -1 & 5 & \vdots & 16 \\ 0 & -2\frac{2}{3} & \frac{1}{3} & \vdots & -\frac{13}{3} \\ 4 & 1\frac{3}{4} & -4\frac{1}{4} & \vdots & -9\frac{1}{4} \end{bmatrix}$$

Replacing $R_3 \to -2\frac{2}{3}R_3 - 1\frac{3}{4}R_2$

$$\begin{bmatrix} 3 & -1 & 5 & \vdots & 16 \\ 0 & -2\frac{2}{3} & \frac{1}{3} & \vdots & -\frac{13}{3} \\ 4 & 0 & 10\frac{1}{4} & \vdots & 32\frac{1}{25} \end{bmatrix}$$

By performing the back substituting in equations, we get

$$10.75z = 32.25$$

$$z = \frac{32.25}{10.75} = 3.0$$

$$-2.667y + 0.333 \times 3 = -4.333$$

$$y = \frac{-4.333 - 1}{-2.667} = \frac{5.333}{2.667} = 2.0$$

$$3x - y + 5z = 16$$

$$3x = 16 + y - 5z = 16 + 2 - 5 \times 3 = 3$$

$$x = \frac{3}{3} = 1$$

Hence, $x = 1, y = 2, z = 3$ **Ans.**

Example 5.31. *Solve the following simultaneous equations by the Gaussian's elimination method.*

$$\begin{aligned} 10x - 7y + 3z + 5w &= 6 \\ -6x + 8y + z - 4w &= 5 \\ 3x + y + 4z - 11w &= 2 \\ 5x - 9y - 2z + 4w &= 7 \end{aligned}$$

Solution:

$$\begin{aligned} 10x - 7y + 3z + 5w &= 6 \\ -6x + 8y + z - 4w &= 5 \\ 3x + y + 4z - 11w &= 2 \\ 5x - 9y - 2z + 4w &= 7 \end{aligned}$$

By writing the augmented coefficient matric in the form of AX = B, we get

$$\begin{bmatrix} 10 & -7 & 3 & 5 \\ -6 & 8 & -1 & 4 \\ 3 & 1 & 4 & 11 \\ 5 & -9 & 0 & 4 \end{bmatrix} \begin{bmatrix} x \\ y \\ z \\ w \end{bmatrix} = \begin{bmatrix} 6 \\ 5 \\ 2 \\ 7 \end{bmatrix}$$

or

$$\begin{bmatrix} 10 & -7 & 3 & 5 & 6 \\ -6 & 8 & -1 & 4 & 5 \\ 3 & 1 & 4 & 11 & 2 \\ 5 & -9 & -2 & 4 & 7 \end{bmatrix}$$

He shall apply the transformations to get lower triangular matrix of zeros elements

$$\begin{bmatrix} 3 & 1 & 4 & 11 & 2 \\ 5 & -9 & -2 & 4 & 7 \\ -6 & 8 & -1 & -4 & 5 \\ 10 & -7 & 3 & 5 & 6 \end{bmatrix}$$

First, do the following transformations:

$R_2 \to R_2 - \frac{5}{3} R_1$; $R_3 \to E_3 + 2R_1$; $R_4 \to R_4 - \frac{10}{3} R_1$ to obtain the following matrix:

$$\begin{bmatrix} 3 & 1 & 4 & 11 & 2 \\ 0 & -32/3 & -26/3 & -43/3 & 11/3 \\ 0 & 10 & 7 & 18 & 9 \\ 0 & -31/3 & -31/3 & -95/3 & -2/3 \end{bmatrix}$$

Second, do the following transformations:

$R_2 \to 3R_2$; $R_4 \to 3R_4$ to eliminate the denominator 3 of rows two and four.

$$\begin{bmatrix} 3 & 1 & 4 & 11 & 2 \\ 0 & -32 & -26 & -43 & 11 \\ 0 & 10 & 7 & 18 & 9 \\ 0 & -31 & -31 & -95 & -2 \end{bmatrix}$$

Third, do the following transformations:

$R_2 \to R_2 - R_4$

$$\begin{bmatrix} 3 & 1 & 4 & 11 & 2 \\ 0 & -1 & -5 & 52 & 13 \\ 0 & 10 & 7 & 18 & 9 \\ 0 & -31 & -31 & -95 & -2 \end{bmatrix}.$$

Fourth, do the following transformations: $R_3 \to R_3 + 10R_2$; $R_4 \to R_4 - 31R_2$

$$\begin{bmatrix} 3 & 1 & 4 & 11 & 2 \\ 0 & -1 & 5 & 52 & 13 \\ 0 & 0 & 57 & 538 & 139 \\ 0 & -31 & -196 & -1707 & -405 \end{bmatrix}$$

Fifth, Replace R_4 by $\frac{1}{3}$ R_4.

$$\begin{bmatrix} 3 & 1 & 4 & 11 & 2 \\ 0 & -1 & 5 & 52 & 13 \\ 0 & 0 & 57 & 538 & 139 \\ 0 & 0 & -62 & -569 & -135 \end{bmatrix}$$

Sixth, Replace R_3 by $(R_3 + R_4)$

$$\begin{bmatrix} 3 & 1 & 4 & 11 & 2 \\ 0 & -1 & 5 & 52 & 13 \\ 0 & 0 & -5 & -31 & 4 \\ 0 & 0 & -62 & -569 & -135 \end{bmatrix}$$

Seventh, Replace R_4 by $R_4 - \frac{62}{5} R_3$,

$$\begin{bmatrix} 3 & 1 & 4 & 11 & 2 \\ 0 & -1 & 5 & 52 & 13 \\ 0 & 0 & -5 & -31 & 4 \\ 0 & 0 & 0 & -923 & -923 \end{bmatrix}$$

Now rewrite the resultant augmented matrix in the form of $Ax = B$

$$\begin{bmatrix} 3 & 1 & 4 & 11 \\ 0 & -1 & 5 & 52 \\ 0 & 0 & -5 & -31 \\ 0 & 0 & 0 & -923 \end{bmatrix} \begin{bmatrix} x \\ y \\ z \\ w \end{bmatrix} = \begin{bmatrix} 2 \\ 13 \\ 4 \\ -923 \end{bmatrix}$$

Now do back ward computation for bottom to top.

i.e., $$-923w = -923$$

or $$w = 1$$

From 3rd row weget

$$-5z - 31w = 4$$

$$-5z - 31 \times 1 = 4$$

$$z = \frac{4 + 31}{-5} = -7$$

From second row (R_2), weget

$$-y + 5z + 52w = 13$$

$$-y + 5 \times (-7) + 52 \times 1 = 13$$

$$-y - 35 + 52 = 13$$

$$y = 4$$

From first row (R_1) weget

$$\begin{aligned} 3x + y + 4z + 11w &= 2 \\ 3x + 4 - 28 + 11 \times 1 &= 2 \\ 3x &= 15 \\ x &= 5 \end{aligned}$$

The solutions of the system of equations is:

$$x = 5 \quad ; \quad y = 4 \quad ; \quad z = -7 \quad ; \quad w = 1 \quad \textbf{Ans.}$$

Example 5.32. *Solve the following simultaneous equations using matrix inversion method.*

$$\begin{aligned} 3x + y + 2z &= 3 \\ 2x - 3y - z &= -3 \\ x + 2y + z &= 4 \end{aligned}$$

Solution: The given equations are:

$$\begin{aligned} 3x + y + 2z &= 3 \\ 2x - 3y - z &= -3 \\ x + 2y + z &= 4 \end{aligned}$$

The given equations may be written in the form of matrix notation:

$$AX = B \qquad ...(i)$$

where $A = \begin{bmatrix} 3 & 1 & 2 \\ 2 & -3 & -1 \\ 1 & 2 & 1 \end{bmatrix} X = \begin{bmatrix} x \\ y \\ z \end{bmatrix} B = \begin{bmatrix} 3 \\ -3 \\ 4 \end{bmatrix}$

The inverse of the matrix may be obtained by finding its adj (A).

The matrix of co-factors of the elements of A are:

$$\begin{bmatrix} -1 & -3 & 7 \\ 3 & 1 & -5 \\ 5 & -7 & -11 \end{bmatrix}$$

The adj (A) in the transpose of matrix of co-factors, *i.e.*,

$$\begin{bmatrix} -1 & 3 & 5 \\ -3 & 1 & -7 \\ 7 & -5 & -11 \end{bmatrix}$$

The value of the determinant of the matrix A

$$= 3(-3 + 2) - 1(2 + 1) + 2(4 + 3) = 8$$

The inverse $\quad A^{-1} = \dfrac{1}{8}\begin{bmatrix} -1 & 3 & 5 \\ -3 & 1 & -7 \\ 7 & -5 & -11 \end{bmatrix}$

Multiplying eqn. (*i*) by A^{-1} on both sides, we get

$$A^{-1} \cdot AX = A^{-1} \cdot B$$

$$X = A^{-1}B \qquad (\because \quad A^{-1} \cdot A = 1)$$

The solution of X may be obtained as under:

$$X = \begin{bmatrix} x \\ y \\ z \end{bmatrix} \frac{1}{8} \begin{bmatrix} -1 & 3 & 5 \\ -3 & 1 & 7 \\ 7 & -5 & -11 \end{bmatrix} \begin{bmatrix} 3 \\ -3 \\ 4 \end{bmatrix}$$

$$X = \frac{1}{8} \begin{bmatrix} 8 \\ 16 \\ -8 \end{bmatrix}$$

$$\begin{bmatrix} x \\ y \\ z \end{bmatrix} = \begin{bmatrix} 1 \\ 2 \\ -1 \end{bmatrix}$$

i.e., $x = 1, y = 2, z = -1$ **Ans.**

Example 5.33. *Solve the following system of simultaneous linear equations using Gauss elimination method.*

$$x + 2y + z + 4w = 13$$
$$2x + 4y + 3w = 28$$
$$4x + 2y + 2z + w = 20$$
$$-3x + y + 3z + 2w = 6$$

Solution: The augmented matrix of the given linear equations, is

$$\begin{bmatrix} 1 & 2 & 1 & 4 \\ 2 & 0 & 4 & 3 \\ 4 & 2 & 2 & 1 \\ -3 & 1 & 3 & 2 \end{bmatrix} \begin{bmatrix} x \\ y \\ z \\ w \end{bmatrix} = \begin{bmatrix} 13 \\ 28 \\ 20 \\ 6 \end{bmatrix}$$

By applying Gauss elimination method and using row transformation, we get a triangularised matrix A, as under

Transform $R_2 \to R_2 - 2R_1$, $R_3 \to R_3 - 4R_1$, $R_4 \to R_4 - 3R_1$ to get

$$\begin{bmatrix} 1 & 2 & 1 & 4 \\ 0 & -4 & 2 & -5 \\ 0 & -6 & -2 & -15 \\ 0 & 7 & 6 & 14 \end{bmatrix} \begin{bmatrix} x \\ y \\ z \\ w \end{bmatrix} = \begin{bmatrix} 13 \\ 2 \\ -32 \\ 45 \end{bmatrix}$$

Transform $R_3 \to R_3 - 1.75R_2$; $R_4 \to R_3 + 1.75R_2$ to get

$$\begin{bmatrix} 1 & 2 & 1 & 4 \\ 0 & -4 & 2 & -5 \\ 0 & 0 & -5 & -7.5 \\ 0 & 0 & 9.5 & 5.25 \end{bmatrix} \begin{bmatrix} x \\ y \\ z \\ w \end{bmatrix} = \begin{bmatrix} 13 \\ 2 \\ -35 \\ 48.5 \end{bmatrix}.$$

Transform $R_4 \to R_4 + 1.9R_3$ to get

$$\begin{bmatrix} 1 & 2 & 1 & 4 \\ 0 & -4 & 2 & -5 \\ 0 & 0 & -5 & -7.5 \\ 0 & 0 & 0 & -9 \end{bmatrix} \begin{bmatrix} x \\ y \\ z \\ w \end{bmatrix} = \begin{bmatrix} 13 \\ 2 \\ -35 \\ -5 \end{bmatrix}$$

By using the back substitutions, we get

$$\begin{aligned} -9w &= -9 & \text{or} \quad w &= 1.0 \\ -6z - 7.5 &= -35 & \text{or} \quad z &= 5.5 \\ -4y + 2z - 5w &= 2 & \text{or} \quad z &= 1.0 \\ x + 2y + z + 4w &= 13 & x &= 1.5 \end{aligned}$$

The solution of the given equations is as under:

$x = 1.5 \; ; \; y = 1 \; ; \; z = 5.5 \; ; \; w = 1$ **Ans.**

Example 5.34. *Solve the following set of simultaneous equations by the Gauss-elimination method:*

$$\begin{aligned} 2x_1 + x_2 + 2x_3 + x_4 &= 6 \\ 6x_2 - 6x_2 + 6x_3 + 12x_4 &= 36 \\ 4x_1 + 3x_2 + 3x_3 - 3x_4 &= -1 \\ 2x_1 + 2x_2 - x_3 + x_4 &= 10 \end{aligned}$$

Solution: The given simultaneous equations can be expressed in matrix form as under:

$$\begin{bmatrix} 2 & 1 & 2 & 1 \\ 6 & -6 & 6 & 12 \\ 4 & 3 & 3 & -3 \\ 2 & 2 & -1 & 1 \end{bmatrix} \begin{bmatrix} x_1 \\ x_2 \\ x_3 \\ x_4 \end{bmatrix} = \begin{bmatrix} 6 \\ 36 \\ -1 \\ 10 \end{bmatrix}$$

$$\left.\begin{array}{l} \text{Replace 2nd row by } R_2 - 3R_1 \\ \text{Replace 3rd row by } R_3 - 2R_1 \\ \text{Replace 4th row by } R_4 - R_1 \end{array}\right\} \text{to get the following matrix}$$

$$\begin{bmatrix} 2 & 1 & 2 & 1 \\ 0 & -9 & 0 & 9 \\ 0 & 1 & -1 & -5 \\ 0 & 1 & -3 & 0 \end{bmatrix} \begin{bmatrix} x_1 \\ x_2 \\ x_3 \\ x_4 \end{bmatrix} = \begin{bmatrix} 6 \\ 18 \\ -13 \\ 4 \end{bmatrix}$$

Replace second row by $\frac{R_2}{-9}$ to get the following:

$$\begin{bmatrix} 2 & 1 & 2 & 1 \\ 0 & 1 & 0 & -1 \\ 0 & 1 & -1 & -5 \\ 0 & 1 & -3 & 0 \end{bmatrix} \begin{bmatrix} x_1 \\ x_2 \\ x_3 \\ x_4 \end{bmatrix} = \begin{bmatrix} 6 \\ -2 \\ -13 \\ 4 \end{bmatrix}$$

Replace third row by $R_3 - R_2$ to get the following

$$\begin{bmatrix} 2 & 1 & 2 & 1 \\ 0 & 1 & 0 & -1 \\ 0 & 0 & -1 & -4 \\ 0 & 0 & -3 & 1 \end{bmatrix} \begin{bmatrix} x_1 \\ x_2 \\ x_3 \\ x_4 \end{bmatrix} = \begin{bmatrix} 6 \\ -2 \\ -11 \\ 6 \end{bmatrix}$$

Replace 4th row by $R_4 - 3R_2$ to get the following

$$\begin{bmatrix} 2 & 1 & 2 & 1 \\ 0 & 1 & 0 & -1 \\ 0 & 0 & -1 & -4 \\ 0 & 0 & 0 & 13 \end{bmatrix} \begin{bmatrix} x_1 \\ x_2 \\ x_3 \\ x_4 \end{bmatrix} = \begin{bmatrix} 6 \\ -2 \\ -11 \\ 39 \end{bmatrix}$$

Now, we get the following equations:

$$2x_1 + x_2 + 2x_3 + x_4 = 6 \quad ...(i)$$

$$x_2 + 0x_3 - x_4 = -2 \quad ...(ii)$$

$$-x_3 + 4x_4 = -11 \quad ...(iii)$$

$$13x_4 = 39 \quad ...(iv)$$

From eqn (*iv*) we get

$$13x_4 = 39 \quad \text{or} \quad x_4 = 3$$

By substituting the value of x_4 in eqn. (*iii*) we get

$$-x_3 - 4 \times 3 = -11 \quad \text{or} \quad x_3 = -1$$

By substituting the value of x_4 in eqn. (*ii*) we get

$$x_2 = -2 + 3 \quad \text{or} \quad x_2 = 1$$

By substituting the value of x_4, x_3, x_2 in eqn. (*i*) we get

$$2x_1 + 1 + 2 + 3 = 6 \quad \text{or} \quad x_1 = 2$$

Here, $x_1 = 2, x_2 = 1, x_3 = -1, x_4 = 3$ **Ans.**

Example 5.35. *Use the Gauss-Siedel algorithm to solve the following set of equations up to three iterations.*

$$\begin{bmatrix} 8 & 2 & 3 \\ 1 & -9 & 2 \\ 2 & 3 & 6 \end{bmatrix} \begin{bmatrix} x_1 \\ x_2 \\ x_3 \end{bmatrix} = \begin{bmatrix} 30 \\ 1 \\ 31 \end{bmatrix}$$

Solution:

$$8x_1 + 2x_2 + 3x_3 = 30 \quad ...(1)$$

$$x_1 - 9x_2 + 2x_3 = 1 \quad ...(2)$$

$$2x_2 + 3x_2 + 6x_3 = 31 \quad ...(3)$$

Putting $x_2 = 0$ and $x_3 = 0$ in eqn. (1), we get

$$8x_1 + 0 + 0 = 30$$

or $$x_1 = \frac{30}{8} = 3.75$$

Putting $x_1 = 3.75$ and $x_3 = 0$ in eqn. (2), we get

$$3.75 - 9x_2 + 0 = 1$$

or $$x_2 = \frac{3.75 - 1}{9} = 0.31$$

Putting $x_1 = 3.75$, and $x_2 = 0.31$ in eqn. (3), we get

$$2 \times 3.75 + 3 \times 0.31 + 6x_3 = 31$$

or $$x_3 = \frac{31 - 3.75 - 0.93}{6} = 3.76$$

This is the end of the first interation.

Putting $x_2 = 0.31$ and $x_3 = 3.76$ in eqn. (1), we get

$$8x_1 + 2 \times 0.31 + 3 \times 3.76 = 30$$

or $$x_1 = \frac{30 - 0.62 - 11.28}{8} = 2.26$$

Putting $x_1 = 2.26$ and $x_3 = 3.76$ in eqn. (2), we get

$$2.26 - 9x_2 + 2 \times 3.76 = 1$$

or $$x_2 = \frac{2.26 + 7.52 - 1}{9} = 0.98$$

Putting $x_1 = 2.26$ and $x_2 = 0.98$ in eqn. (3), we get

$$2 \times 2.26 + 3 \times 0.98 + 6x_3 = 31$$

or $$x_3 = \frac{31 - 4.52 - 2.94}{6} = 3.92$$

This is the end of the second interation.

Putting $x_2 = 0.98$ and $x_3 = 3.92$ in eqn. (1), we get

$$8x_1 + 2 \times 0.98 + 3 \times 3.92 = 30$$

or $$x_1 = \frac{30 - 1.96 - 11.76}{8} = 2.04$$

Putting $x_1 = 2.04$ and $x_3 = 3.92$ in eqn. (2), we get

$$2.04 - 9x_2 + 2 \times 3.92 = 1$$

or $$x_2 = \frac{2.04 + 7.84 - 1}{9} = 0.99$$

Putting $x_1 = 2.04$ and $x_2 = 0.99$ in eqn. (3), we get

$$2 \times 2.04 + 3 \times 0.99 + 6x_3 = 31$$

or $$x_3 = \frac{31 - 4.08 - 2.97}{6} = 3.99$$

This is the end of the third interation.

Putting $x_2 = 0.99$ and $x_3 = 3.99$ in eqn. (1), we get

$$8x_1 + 2 \times 0.99 + 3 \times 3.99 = 30$$

or $$x_1 = \frac{31 - 1.98 - 11.97}{8} = 2.0$$

Putting $x_1 = 2$ and $x_3 = 3.99$ in eqn. (2), we get

$$2 - 9x_2 + 2 \times 3.99 = 1$$

or $$x_2 = \frac{2 + 7.98 - 1}{9} = 1.0$$

Putting $x_1 = 2$, $x_2 = 1$ in eqn. (3), we get

$$2 \times 2 + 3 \times 1 + 6x_3 = 31$$

or $$x_3 = \frac{31 - 4 - 3}{6} = 4$$

$$\left.\begin{array}{l} x = 1 \\ y = 1 \\ z = 1 \end{array}\right\} \quad \textbf{Ans.}$$

5.14 DIFFERENTIAL EQUATIONS OF FIRST ORDER

There are many methods for the solution of differential equations of the first order, each provides solution of varying accuracy.

Jacobi's iteration method.

Consider the following equations:

$$\left.\begin{array}{l} a_1 x + b_1 y + c_1 z = d_1 \\ a_2 x + b_2 y + c_2 z = d_2 \\ a_3 x + b_3 y + c_3 z = d_3 \end{array}\right\} \qquad ...(1)$$

These equations can be written as under:

$$\left.\begin{aligned} x &= \frac{1}{a_1}(d_1 - b_1 y - c_1 z) \\ y &= \frac{1}{b_2}(d_2 - a_2 x - c_2 z) \\ z &= \frac{1}{c_3}(d_3 - a_3 x - b_3 y) \end{aligned}\right\} \quad \ldots(2)$$

Let the initial approximations be x_0, y_0 z_0 for th values of x, y, z. By substituting these values on the right sides of (1) the first approximations are given by

$$x_1 = \frac{1}{a_1}(d_1 - b_1 y_0 - c_1 z_0)$$

$$y_1 = \frac{1}{b_2}(d_2 - a_2 x_0 - c_2 z_0)$$

$$z_1 = \frac{1}{c_3}(d_3 - a_3 x_0 - b_3 y_0)$$

By substituting the values x_1, y_1, z_1 on the right sides of eqn. (2) the second approximations are given by

$$x_2 = \frac{1}{a_1}(d_1 - b_1 y_1 - c_1 z_1)$$

$$y_2 = \frac{1}{b_2}(d_2 - a_2 x_1 - c_2 z_1)$$

$$z_1 = \frac{1}{c_3}(d_3 - a_3 x_1 - b_2 y_1)$$

This process is repeated till the difference between two successive approximations is negligible.

Example 5.36. *Solve by Jacobis iteration method, the equations*

$$\begin{aligned} 20x + y - 2z &= 16 \\ 3x + 20y - z &= -20 \\ 2x - 3y + 20z &= 24 \end{aligned}$$

Solution: The given equations may be written as under:

$$\left.\begin{aligned} x &= \frac{1}{20}(16 - y + 2z) \\ y &= \frac{1}{20}(-20 - 3x + z) \\ z &= \frac{1}{20}(24 - 9x + 3y) \end{aligned}\right\} \quad \ldots(1)$$

Let as assume that initial approximations are:

$$x = 0 \;;\; y = 0 \;;\; z = 0$$

By substituting these values on the right sides of equations (1), we get

$$x_1 = \frac{1}{20}(16 - 0 + 0) = 0.8$$

$$y_1 = \frac{1}{20}(-20 - 0 + 0) = -1.0$$

$$z_1 = \frac{1}{20}(24 - 0 + 0) = 1.2$$

By substituting these values on the right of equations (1), we get

$$x_2 = \frac{1}{20}(16 + 1.0 + 2 \times 1.2) = 0.97$$

$$y_2 = \frac{1}{20}(-20 - 2.4 + 1.2) = -1.06$$

$$z_2 = \frac{1}{20}(24 - 1.6 - 3) = 0.97$$

By substituting these values on the right of equations (1) we get

$$x_3 = \frac{1}{20}(16 + 1.06 + 1.94) = 0.95$$

$$y_3 = \frac{1}{20}(-20 - 2.91 + 0.97) = -1.097$$

$$z_3 = \frac{1}{20}(24 - 1.94 - 3.18) = 0.944$$

By substituting these values on the right of equations (1) we get

$$x_4 = \frac{1}{20}(16 + 1.097 + 1.888) = 0.95$$

$$y_4 = \frac{1}{20}(-20 - 2.85 + 0.944) = 1.095$$

$$z_4 = \frac{1}{20}(24 - 1.90 - 3.291) = 0.940$$

By substituting these values on the right of equations (1), we get

$$x_5 = \frac{1}{20}(16 + 1.095 + 1.880) = 0.94875$$

$$y_5 = \frac{1}{20}(-20 - 2.85 + 0.940) = 1.0955$$

$$z_5 = \frac{1}{20}(24 - 1.90 - 3.285) = 0.94075$$

The values in the 4th and 5th iterations being practically same, the solution of the equations is

$$x = 0.95; \quad -1.1; \quad 0.94$$

Check: (*i*) $20 \times 0.95 - 1.1 - 2 \times 0.94 = 19 - 1.1 - 1.88 = 16.02$
(*ii*) $3 \times 0.95 + 20 \times (-1.1) - 0.94 = 2.85 - 22 - 0.94 = -20.09$ **Ans.**
(*iii*) $2 \times 0.95 - 3 \times (-1.1) + 20 \times 0.94 = 1.9 + 3.3 + 1.88 = 24.00$

5.15 SOLUTION OF DIFFERENTIAL EQUATION BY EULER'S METHOD

The differential equation of the first order, may be written as under:

$$\frac{dy}{dx} = f(x, y) \quad \text{...(1)}$$

Integrating the equation (1), we get

$$y = f(x) \quad \text{...(2)}$$

Now $y = f(x)$ represents a smooth curve in the X-Y plane. The curve $y = f(x)$ may be assumed as composed of a large number of straight lines, each of short length.

Referring to Fig. 5.9 we derive an approximate relationship for the increment in y for $x = x_1$.

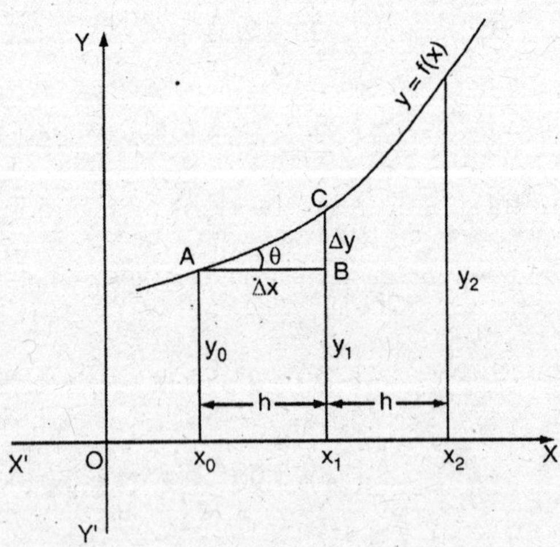

Fig. 5.9

Let the ordinate y intersect the curve $y = f(x)$ at A.

Draw AB parallel to X-axis. From the right angled triangle ABC, we get

$$\tan\theta = \frac{\Delta y}{\Delta x} \text{ where } \angle CAB = \theta.$$

$\therefore \quad \Delta y = \Delta x \tan \theta = \left(\frac{dy}{dx}\right)_0 \cdot \Delta x$, where $\frac{dy}{dx}$ is the slope of the curve at A.

But, $$y_1 = y_0 + \Delta y$$

$$\therefore \quad y_1 = y_0 + \left(\frac{dy}{dx}\right)_0 \Delta x$$

Similarly, for $x = x_2$ i.e. $(x_1 + \Delta x)$ where Δx is an equal interval between ordinates.

$$y_2 = y_1 + \left(\frac{dy}{dx}\right)_1 . \Delta x$$

$$y_3 = y_2 + \left(\frac{dy}{dx}\right)_2 \Delta x$$

....

....

$$y_n = y_{n-1} + \left(\frac{dy}{dx}\right)_{n-1} \Delta x \quad \text{...(3)}$$

By taking Δx small enough, we determine the integral of (1) as a set of corresponding values of x and y.

It may however be noted that the numerical solution by this method gives a poor approximation to the true value.

For obtaining better results, Euler's modified method discussed later may be used.

Example 5.37. *Given:* $\frac{dy}{dx} = 3x^2 + 2y, y(0) = 1.$

Using Euler's method, obtain y at $x = 0.5, 1.0, 1.5$.

Solution: Here $x_0 = 0, y_0 = 1, h = 0.5$

(*i*) $x = 0.5$

Substituting the values in eqn. (3)

$$\begin{aligned} y_1 &= y_0 + \left(\frac{dy}{dx}\right)_0 \Delta x \\ &= y_0 + h f(x_0, y_0) \\ &= 1 + 0.5 \times (3 \times 0^2 + 2 \times 1) \\ y_1 &= 1 + 1 = 2 \end{aligned}$$

(*ii*) $x = 1.0$

$$y_2 = y_1 + \left(\frac{dy}{dx}\right)_1 \cdot \Delta x$$

$$= y_1 + h\ f(x_0, y_0)$$

$$= 2 + 0.5 \times (3 \times 1^2 + 2 \times 2),$$

using the value of $y = 2$ obtained in step (*i*)

$$= 2 + 0.5\ (3 + 4)$$

$$y_2 = 2 + 3.5 = 5.5$$

(*iii*) $x = 1.5$

$$y_3 = y_2 + \left(\frac{dy}{dx}\right)_2 \cdot \Delta x$$

$$= y_2 + h\ f(x_0, y_0)$$

$$= 5.5 + 0.5 \times (3 \times 1.5^2 + 2 \times 5.5),$$

using the value of $y = 5.5$ obtained in step (*ii*)

$$= 5.5 + 0.5 \times 17.75 = 14.375$$

$$\left.\begin{array}{ll} x = 0.5 & y = 2 \\ x = 1.0 & y = 5.5 \\ x = 1.5 & y = 14.375 \end{array}\right\} \textbf{Ans.}$$

Example 5.38. *Using Euler's method, solve numerically the following equation.*

$$y' = y^2 - \frac{y}{x},\ y(1) = 1$$

for the interval 1, (0.1), 1.5.

Solution: $x = 1, y = 1, h = 0.1$

$$y_{1.1} = y_1 + h\ f(x_{1.1}, y_{1.1})$$

$$= 1 + 0.1\left(1^2 - \frac{1}{1.1}\right)$$

$$= 1 + 0.0091 = 1.0091$$

$$y_{1.2} = y_{1.1} + h\ f\left(y_{1.1}^2 - \frac{y_{1.1}}{x_{1.2}}\right)$$

$$= 1.0091 + 0.1\left((1.0091)^2 - \frac{1.0091}{1.2}\right)$$

$$= 1.0091 + 0.1\ (0.1777366) = 1.0268$$

$$y_{1.3} = y_{1.2} + h\ f\left({}^{2}_{1.2} - \frac{y_{1.2}}{y_{1.3}}\right)$$

$$= 1.0268 + 0.1\left(1.0268^2 - \frac{1.0268}{1.3}\right)$$

$$= 1.028 + 0.0265 = 1.0533$$

$$y_{1.4} = y_{1.3} + h\left(y_{1.3}^2 - \frac{y_{1.3}}{x_{1.4}}\right)$$

$$= 1.0533 + 0.1\left(1.0533^2 - \frac{1.0533}{1.4}\right)$$

$$= 1.0533 + 0.0357 = 1.089$$

$$y_{1.5} = y_{1.4} + h\int\left(y_{1.4}^2 - \frac{y_{1.4}}{x_{1.5}}\right)$$

$$= 1.089 + 0.1\left(1.089^2 - \frac{1.089}{1.5}\right)$$

$$= 1.089 + 0.0460 = 1.135$$

x	1	1.1	1.2	1.3	1.4	1.5	
y	1	1.0091	1.0268	1.0533	1.089	1.135	**Ans.**

5.16 EULER'S MODIFIED METHOD

This method is used to start the solution of the differential equations where no other method is found suitable.

Let $$\frac{dy}{dx} = f(x, y) \qquad ...(i)$$

and h is the assumed interval.

(A) The value of y_1 may be determined as under:

(*i*) With the help of the given value of y_0, calculate the approximate value of y_1, *i.e.*,

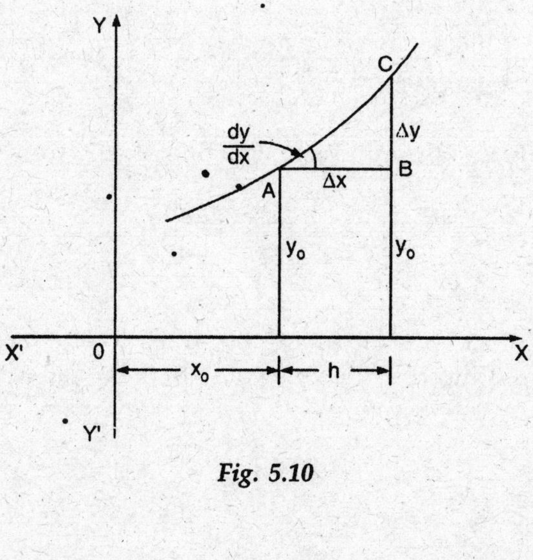

Fig. 5.10

$$y_1^{(1)} = y_0 + \Delta y$$

or
$$y_1^{(1)} = y_0 + \left(\frac{dy}{dx}\right)_0 h.$$

(*ii*) Substituting the value of $y_1^{(1)}$ in the given differential eqn. (1), calculate the approximate value of $\frac{dy}{dx}$ at the end of the first interval, *i.e.*,

$$\left(\frac{dy}{dx}\right)_1^{(1)} = \left(fx_1, y_1^{(1)}\right)$$

(*iii*) Taking the average value of $\frac{dy}{dx}$ at the ends of the intervals x_0 to x_1 and multiplying the same by the assumed interval 'h', we get the next approximate value of Δy, *i.e.*,

$$\Delta y \approx \frac{\left(\frac{dy}{dx}\right)_0 + \left(\frac{dy}{dx}\right)_1^{(1)}}{2} h.$$

(*iv*) Now, calculate the second approximate value of y_1, *i.e.*,

$$y_1^{(2)} = \frac{\left(\frac{dy}{dx}\right)_0 + \left(\frac{dy}{dx}\right)_1^{(1)}}{2} h$$

(*v*) Calculate the second appriximate value for $\left(\frac{dy}{dx}\right)_1^{(2)}$ by substituting the value in the differential equation *i.e.*,

$$\left(\frac{dy}{dx}\right)_1^{(2)} = \int(x_1 y_1^{(2)})$$

(*vi*) Calculate the 3rd approximate value of $y_1^{(3)}$ from the formula

$$y_1^{(3)} = y_0 + \frac{\left(\frac{dy}{dx}\right)_0 + \left(\frac{dy}{dx}\right)_1^{(2)}}{2} h$$

Repeat the process for obtaining the 4th, 5th, 6th, ... n^{th} value of y_1 till no change is produced in the value of y_1 to the required number of digits.

(B) The value of y_2 may be determined as under:

The value of y_2 for the next intervals x_1 to x_2 or $(x_1 + h)$ may be determined exactly in the same manner as explained for the value of y_1. It means we first find the approximate value of $\Delta y = \left(\frac{dy}{dx}\right)_1 h$ and then get the approximate value of $y_2^{(1)} = y_1 + \left(\frac{dy}{dx}\right)_1 h$ and so on.

Example 5.39. *Calculate the few values of y for the differential equation*

$$\frac{dy}{dx} = x + y$$

with the initial condition $x_0 = 0, y_0 = 1.$

Solution: The given differential equation is

$$\frac{dy}{dx} = x + y \qquad ...(i)$$

Substituting the initial value of $x_0 = 0$ and $y_0 = 1$, we get

$$\left(\frac{dy}{dx}\right)_0 = 0 + 1 = 1$$

Assuming the interval $h = 0.05$, we get

$$y_1^{(1)} = y_0 + \left(\frac{dy}{dx}\right)_0 h$$

$$= 1 + 1 \times 0.05 = 1.05$$

Now, calculate the value of $\left(\frac{dy}{dx}\right)_1$ by substituitng the value $x_1 = 0.05$ and $y_1^{(1)} = 1.05$, *i.e.*,

$$\left(\frac{dy}{dx}\right)_1^{(1)} = 0.05 + 1.05 = 1.10$$

$$\therefore \qquad y_1^{(2)} = y_0 + \frac{\left(\frac{dy}{dx}\right)_0 + \left(\frac{dy}{dx}\right)_1^{(1)}}{2} h$$

$$= 1 + \frac{1 + 1.10}{2} \times 0.05 = 1.0525$$

Calculate the second approximation to $\left(\frac{dy}{dx}\right)_1^{(2)}$

$$\left(\frac{dy}{dx}\right)_1^{(2)} = 0.05 + 1.0525 = 1.1025$$

The third approximation to y_1

$$y_1^{(3)} = y_0 \frac{\left(\frac{dy}{dx}\right)_0 + \left(\frac{dy}{dx}\right)_1^{(2)}}{2} h$$

or
$$y_1^{(3)} = 1 + \frac{1 + 1.1025}{2} \times 0.05 = 1.05256$$

Then, third approximation to $\left(\frac{dy}{dx}\right)_1^{(3)}$

$$\left(\frac{dy}{dx}\right)_1^{(3)} = 0.05 + 1.05256 = 1.10256$$

The fourth approximation to y_1

$$y_1^{(4)} = y_0 + \frac{\left(\frac{dy}{dx}\right)_0 + \left(\frac{dy}{dx}\right)_1^{(3)}}{2} \times h$$

$$= 1 + \frac{1 + 1.10256}{2} \times 0.05 = 1.05256$$

As $y_1^{(3)}$ and $y_1^{(4)}$ are the same, no more approximation is needed and the value of $y_1 = 1.05256$

and
$$\left(\frac{dy}{dx}\right)_1 = 1.10256 \text{ may be accepted.}$$

To obtain the first apprimation to y, we use the following equation

$$y_{n+1}^{(1)} + 2hyn = y_{n-1} + 2hy'_n$$

$$y_2 = y_0 + 2h\left(\frac{dy}{dx}\right)_1$$

i.e.,
$$y_2^{(1)} = 1 + 2 \times 0.05 \times (1.10256)$$
$$= 1.11026$$

$\therefore$
$$\left(\frac{dy}{dx}\right)_2^{(1)} = (0.05 + 0.05) + 1.11026$$

$$(\because \quad x_2 = h + h = 2h = -1)$$

$$= 1.21026$$

$$y_2^{(2)} = y_1 + \frac{\left(\frac{dy}{dx}\right)_1 + \left(\frac{dy}{dx}\right)_2^1}{2} \times h$$

$$= 1.05256 + \frac{1.05256 + 1.21026}{2} \times 0.05$$

$$= 1.05256 + 0.0578205 = 1.11038$$

$$\left(\frac{dy}{dx}\right)_3^2 = 0.1 + 1.11038 = 1.21038$$

$$y_2^{(3)} = 1.05256 + \frac{1.10256 + 1.21038}{2} \times 0.05$$

$$= 1.11038$$

As $y_2^{(3)}$ and $y_2^{(2)}$ are the same we accept

$$y_2 = 1.11038 \text{ and } \left(\frac{dy}{dx}\right)_2 = 1.21038$$

The final results are tabulated here under

x	y	dy/dx	
0.00	1.00000	1.00000	
0.05	1.05256	1.10256	**Ans.**
0.10	1.11038	1.21038	

5.17 BY USING TAYLOR'S SERIES

Let the given function be $y = f(x)$

We may write the Taylor's formula as under:

$$f(x) = f(x_0) + f'(x_0)(x - x_0) + \frac{f'''(x_0)}{2!}(x - x_0)^2 + \frac{f'''(x_0)}{3!}(x - x_0)^3 \quad ...(1)$$

$$y = y_0 + y'_0(x - x_0) + \frac{y''_0}{2!}(x - x_0)^2 + \frac{y'''_0}{3!}(x - x_0)^3 + \frac{y''''_0}{4!}(x - x_0)^4 + ... \quad ...(2)$$

where x_0 and y_0 are the initial values of x and y respectively.

Now, compute the values of y's both to the right and to the left of the point $x = x_0$

Substituting the following in eqn. (2)

$$y_1 = \int(x_0 + h)$$

$$y_2 = \int(x_0 + 2h)$$

$$y_{-1} = \int(x_0 + h)$$

$$y_{-2} = \int(x_0 - 2h)$$

$$y_1 = y_0 + y'_0 h + \frac{y_0''h^2}{2} + \frac{y_0'''h_3}{3!} + \frac{y_0''''h^4}{4!} + \frac{y_0'''''h^5}{5!} + \qquad ...(3)$$

$$y_2 = y_0 + y_0'(2h) + \frac{y_0''(2h)^2}{2} + \frac{y_0'''(2h)^3}{3!} + \frac{y_0''''(2h)^4}{4!} + \frac{y_0'''''(2h)^5}{5!} \qquad ...(4)$$

$$y_{-1} = y_0 - y'_0(2h) + \frac{y_0''h^2}{2} - \frac{y_0'''h^3}{3!} + \frac{y_0''''h^4}{4!} + \frac{y_0'''''h^5}{5!} + \qquad ...(5)$$

$$y_{-2} = y_0 + y'_0(2h) + \frac{y_0''(2h)^2}{2} - \frac{y_0'''(2h)^3}{3!} + \frac{y_0''''h^4}{4!} + \frac{y_0'''''h^5}{5!} + \qquad ...(6)$$

where h is the interval between the equidistant values of x. If the successive direvatives of the given differential equation $y' = (x, y)$ are easily found, the consecutive five values of y' can be found with the help of initial conditions, the given equation and the eqns (3) (4) (5) and (6).

Example 5.40. *Calculate the five values for the differential equation.*

$$\frac{dy}{dx} = x + y \quad if \quad x_0 = 0, y_0 = 1.$$

Solution: Given:

$$y' = x + y$$
$$y'' = 1 + y'$$
$$y''' = y''$$
$$y'''' = y'''$$
$$y''''' = y''''$$

$$\therefore \quad y'_0 = x_0 + y_0 = 1$$
$$y_0'' = 1 + y' = 1 + 1 = 2$$
$$y_0''' = y'' = 1 + y' = 1 + 1 = 2$$
$$y_0'''' = y''' = 2$$
$$y_0'''' = y''' = 2$$

Assuming $h = 0.05$ and substituting the value of h in eqns. (3), (4) (5) and (6), we get

$$y_0 + y'_0 h + \frac{y'_0''h^2}{2} + \frac{y_0'''h^3}{3!} + ... +$$

$$y_1 = 1 + 1 \times 0.05 + \frac{2\times(.05)}{2} + \frac{2\times(0.05)^3}{3!} + \frac{2\times(0.05)^4}{4!} + \frac{2\times(0.005)^5}{5!} + - +$$

$$= 1 + 0.05 + .0025 + \frac{2\times .000125}{3\times 2} + \frac{2\times .00000625}{4\times 3\times 2} + \frac{2\times .0000003125}{5\times 4\times 3\times 2\times 1}$$

$$= 1 + 0.05 + 0.0025 + 0.000042$$
$$= 1.05254$$

$$y_2 = y_0 + y_0'(2h) + \frac{y_0''(2h)^2}{2} + \frac{y_0'''(2h)^3}{3!} + \frac{y_0''''(2h)^4}{4!} + \frac{y_0'''''(2h)^5}{5!}$$

$$= 1 + 1\times .1 + \frac{2\times(.1)^2}{2} + \frac{2\times(.1)^3}{3!} + \frac{2\times(.1)^4}{4!} + \frac{2\times(.1)^5}{5!}$$

$$= 1 + .1 + .01 + \frac{.001}{3} + \frac{2\times .0001}{4\times 3\times 2} + \ldots$$

$$= 1 + .1 + .01 + .0003 + .000008 + \ldots$$
$$= 1.1103$$

$$y_{-1} = y_0 - y'_0 h + \frac{y_0''h^2}{2} - \frac{y_0'''h^3}{3!} + \frac{y_0''''h^4}{4!} - \frac{y_0'''''h^5}{5!}$$

$$= 1 - 1.05 + \frac{2\times(0.05)^2}{2} - \frac{2\times(.05)^3}{3!} - \frac{2\times(.05)^4}{4!} - \frac{2\times(0.05)^5}{5!} + -$$

$$= 1 - .05 + 0025 - .000042$$
$$= .95246$$

$$y_{-2} = y_0 - y_0'(2h) + \frac{y_0''(2h)^2}{2} - \frac{y_0'''(2h)^3}{3!} - \frac{y_0''''(2h)^4}{4!} - \frac{y_0'''''(2h)^5}{5!} +$$

$$= 1 - .1 + .01 - .0003 + .000008$$
$$= .99998$$

Now, five desired connective values of y' are:

$$y'_{-2} = x_{-2} + y_{-2} = -0.1 + 0.99998 = 0.89998$$
$$y'_{-1} = x_{-1} + y_{-1} = -0.05 + 0.95246 = 0.90246$$
$$y'_0 = x_0 + y_0 = 1$$
$$y'_1 = x_1 + y_1 = .05 + 1.05254 = 1.10254$$
$$y'_2 = x_2 + y_2 = .1 + 1.1103 = 1.2103$$

i.e., 0.89998, 0.90246, 1, 1.10254, 1.2103. **Ans.**

5.18 PICARD'S METHOD OF SUCCESSIVE APPROXIMATION

Let

$\frac{dy}{dx} = f(x, y)$ be a differential equation of the first order.

$$y = y_0 \text{ when } x = x_0 \text{ is the initial condition}$$

then
$$dy = f(x, y)\, dx = \left(\frac{dy}{dx}\right) . dx \qquad ...(i)$$

Integrating the equation between the corresponding limits for x and y, we get

$$\int_{y_0}^{y} dy = \int_{x_0}^{x} f(x, y)\, dx = \int_{x_0}^{x} \left(\frac{dy}{dx}\right) dx$$

or
$$y - y_0 = \int_{x_0}^{x} f(x, y)\, dx = \int_{x_0}^{x} \left(\frac{dy}{dx}\right) dx$$

$$y = y_0 + \int_{x_0}^{x} (x, y)\, dx = y_0 + \int_{x_0}^{x} \left(\frac{dy}{dx}\right) dx \qquad ...(ii)$$

The integral term in the right hand side of the eqn. (*ii*) represents the increment in y by an increment $(x - x_0)$ in x.

If we consider the first term of eqn. (*i*) we find very difficult to solve the equation due to the presence of y under the integral sign as well as outside it. Such an equation is called *integral equation* and can be solved by a process of successive approximation or interations provided the integrations can be performed in the successive steps.

To solve the eqn. (*i*) By Picard's method of successive approximation proceed as under:

(*i*) Find the first approximation for y by substituting y_0 for y in $f(x, y)$ the integrand in eqn. (*ii*), *i.e.*,

$$y^{(1)} = y_0 + \int_{x_0}^{x} f(x, y)\, dx \qquad ....(iii)$$

(*ii*) As the integrand eqn. (*iii*) is now a function of x, the required integration can be performed to determine the value of $y^{(1)}$.

(*iii*) Substitute the first approximation of y in the integrand of eqn. (*ii*) and integrate again to get the second approximation, *i.e.*,

$$y^{(2)} = y_0 + \int_{x_0}^{x} f(x, y^{(1)})\, dx$$

(*iv*) Continue the process successively till desired degree of accuracy is achieved, *i.e.*,

$$y^{(n)} = y_0 + \int_{x_0}^{x} f(x, y^{(n-1)})\, dx \qquad \text{...(A)}$$

Example 5.41. *Tabulate in 0, (.1), .5, the solution of* $\frac{dy}{dx} = 1 + x$, *which passes through (0.1).*

Use Picard's method and obtain answers correct to third decimal place.

Solution: Given: $x = 0; \; y_0 = 1$

From eqn (A), we get

For $n = 1$

$$y^{(1)} = y_0 + \int_{x_0}^{x} (1 + xy_0)\, dx$$

$$= 1 + \int_0^x (1 + x \times 1)\, dx$$

$$= 1 + \int_0^x (1 + x)\, dx$$

$$= 1 + x + \frac{x^2}{2}$$

$$\therefore \quad y^{(1)} = \frac{x^2}{2} + x + 1 \qquad \text{...}(i)$$

$$y^{(2)} = y_0 + \int_{x_0}^{x} (1 + xy^{(1)})\, dx$$

For $n = 2$

$$y^{(2)} = y_0 + \int_{x_0}^{x} \left[1 + x\left(\frac{x^2}{2} + x + 1\right)\right] dx$$

$$= 1 + \int_{x_0}^{x} \left(1 + \frac{x^3}{2} + x^2 + x\right) dx$$

$$= 1 + \int_{x_0}^{x} \left(\frac{x^3}{2} + x^2 + x + 1\right) dx$$

$$= 1 + \frac{x^4}{8} + \frac{x^3}{3} + \frac{x^2}{2} + x$$

$$\therefore \quad y^{(2)} = \frac{x^4}{8} + \frac{x^3}{3} + \frac{x^2}{2} + x + 1 \qquad \text{...}(ii)$$

For $n = 3$

$$y^{(3)} = y_0 + \int_{x_0}^{x} (1 + x \cdot y^{(2)})\, dx$$

$$= 1 + \int_{x_0}^{x} \left[1 + x\left(\frac{x^4}{8} + \frac{x^3}{3} + \frac{x^2}{2} + x + 1\right)\right] dx$$

$$= 1 + \int_{x_0}^{x} \left[1 + \left(\frac{x^5}{8} + \frac{x^4}{3} + \frac{x^3}{2} + x^2 + x\right)\right] dx$$

$$= 1 + \int_{x_0}^{x} \left[\left(\frac{x^5}{8} + \frac{x^4}{3} + \frac{x^3}{2} + x^2 + x + 1\right)\right] dx$$

$$= 1 + \frac{x^6}{48} + \frac{x^5}{15} + \frac{x^4}{8} + \frac{x^3}{3} + \frac{x^2}{2} + x$$

$$= \frac{x^6}{48} + \frac{x^5}{15} + \frac{x^4}{8} + \frac{x^3}{3} + \frac{x^2}{2} + x + 1 \qquad ...(iii)$$

For $n = 4$ $\quad y^{(4)} = y_0 + \int_{x_0}^{x} (1 + xy^{(3)})\, dx$

$$= 1 + \int_{x_0}^{x} \left[1 + x\left(\frac{x^6}{48} + \frac{x^5}{15} + \frac{x^4}{8} + \frac{x^3}{3} + \frac{x^2}{2} + x + 1\right)\right] dx$$

$$= 1 + \int_{x_0}^{x} \left(1 + \frac{x^7}{48} + \frac{x^6}{15} + \frac{x^5}{8} + \frac{x^4}{3} + \frac{x^3}{2} + x^2 + x\right) dx$$

$$y^{(4)} = 1 + \int_{x_0}^{x} \left(\frac{x^7}{48} + \frac{x^6}{15} + \frac{x^5}{8} + \frac{x^4}{3} + \frac{x^3}{2} + x^2 + x + 1\right) dx$$

$$= 1 + \frac{x^8}{48 \times 8} + \frac{x^7}{105} + \frac{x^6}{48} + \frac{x^5}{15} + \frac{x^4}{8} + \frac{x^3}{3} + \frac{x^2}{2} + x$$

$$= \frac{x^8}{384} + \frac{x^7}{105} + \frac{x^6}{48} + \frac{x^5}{15} + \frac{x^4}{8} + \frac{x^3}{3} + \frac{x^2}{2} + x + 1 \qquad ...(iv)$$

For $n = 5$ $\quad y^{(5)} = y_0 + \int_{x_0}^{x} (1 + xy^{(4)})\, dx$

$$= 1 + \int_{x_0}^{x} \left[1 + x\left(\frac{x^8}{384} + \frac{x^7}{105} + \frac{x^6}{48} + \frac{x^5}{15} + \frac{x^4}{8} + \frac{x^3}{3} + \frac{x^2}{2} + x + 1\right)\right] dx$$

$$= 1 + \int_{x_0}^{x} \left(\frac{x^9}{384} + \frac{x^8}{105} + \frac{x^7}{48} + \frac{x^6}{15} + \frac{x^5}{8} + \frac{x^4}{3} + \frac{x^3}{2} + x^2 + x + 1\right) dx$$

$$= 1 + \frac{x^{10}}{3840} + \frac{x^9}{945} + \frac{x^8}{384} + \frac{x^7}{105} + \frac{x^6}{48} + \frac{x^5}{15} + \frac{x^4}{8} + \frac{x^3}{3} + \frac{x^2}{2} + x$$

$$= \frac{x^{10}}{3840} + \frac{x^9}{945} + \frac{x^8}{384} + \frac{x^7}{105} + \frac{x^6}{48} + \frac{x^5}{15} + \frac{x^4}{8} + \frac{x^3}{3} + \frac{x^2}{2} + x + 1 \qquad ...(v)$$

For $n = 6$ $\qquad y^{(6)} = y_0 + \int_{x_0}^{x} (1 + xy^{(5)})\, dx$

$$= 1 + \int_{x_0}^{x} \left[1 + x\left(\frac{x^{10}}{3840} + \frac{x^9}{945} + \frac{x^8}{384} + \frac{x^7}{105} + \frac{x^6}{48} + \frac{x^5}{15} + \frac{x^4}{8} + \frac{x^3}{3} + \frac{x^2}{2} + x + 1\right)\right] dx$$

$$= 1 + \int_{x_0}^{x} \left(\frac{x^{11}}{3840} + \frac{x^{10}}{945} + \frac{x^9}{384} + \frac{x^8}{105} + \frac{x^7}{48} + \frac{x^6}{15} + \frac{x^5}{8} + \frac{x^4}{3} + \frac{x^3}{2} + x^2 + x + 1\right) dx$$

$$= \frac{x^{12}}{3840 \times 12} + \frac{x^{11}}{945 \times 11} + \frac{x^{10}}{3840} + \frac{x^9}{945} + \frac{x^8}{384} + \frac{x^7}{105}$$

$$+ \frac{x^6}{48} + \frac{x^5}{15} + \frac{x^4}{8} + \frac{x^3}{3} + \frac{x^2}{2} + x + 1 \quad \text{...}(vi)$$

Using the result (*i*), (*ii*), (*iii*), (*iv*), (*v*) and (*vi*), calculate the successive approximations for y when $x = 0, 0.1\ 0.2, 0.3, 0.4, 0.5$, to obtain the numerical solutions correct upto 3 decimal places.

(*a*) $x = 0, y = 1$

(*b*) $x = 0.1$

$$y^{(1)} = \frac{(.1)^2}{2} + 0.1 + 1 = 1.105$$

$$y^{(2)} = \frac{(.1)^4}{8} + \frac{(.1)^3}{3} + \frac{(.1)^2}{2} + .1 + 1 = 1.105$$

(*c*) $x = 0.2$,

$$y^{(1)} = \frac{(.2)^4}{8} + .2 + 1 = 1.22$$

$$y^{(2)} = \frac{(.2)^4}{8} + \frac{(.3)^3}{3} + \frac{(.2)^2}{2} + .2 + 1$$

$$= 0.0002 + 0.0027 + 0.02 + 0.2 = 1.223$$

$$y^{(2)} = \frac{(.2)^6}{48} + \frac{(.2)^5}{15} + \frac{(.2)^4}{8} + \frac{(.3)^3}{3} + \frac{(.2)^2}{2} + .2 + 1$$

$$= 0 + 0 + 0 + 0.003 + 0.02 + 0.2 + 1 = 1.223$$

(*d*) $x = 0.3$

$$y^{(1)} = \frac{(.3)^2}{2} + .3 + 1 = 1.345$$

$$y^{(2)} = \frac{(.3)^4}{8} + \frac{(.3)^3}{3} + \frac{(.3)^2}{2} + .3 + 1$$

$$= 0.0010 + 0.0097 + 0.045 + 0.03 + 1 = 1.355$$

$$y^{(3)} = \frac{(.3)^6}{48} + \frac{(.3)^5}{15} + \frac{(.3)^4}{8} + \frac{(.3)^3}{3} + \frac{(.3)^2}{2} + .3 + 1$$

$$= 0 + 0 + 0.001 + 0.009 + 0.045 + .3 + 1 = 1.35$$

(e) $x = 0.4$

$$y^{(1)} = \frac{(.4)^2}{2} + .4 + 1$$

$$= 0.08 + 0.4 + 1 = 1.480$$

$$y^{(2)} = \frac{(.4)^4}{8} + \frac{(.4)^3}{3} + \frac{(.4)^2}{2} + .4 + 1$$

$$= 0.0032 + 0.0213 + 0.08 + 0.4 + 1 = 1.504$$

$$y^{(3)} = \frac{(.4)^6}{48} + \frac{(.4)^5}{15} + \frac{(.4)^4}{8} + \frac{(.4)^3}{3} + \frac{(.4)^2}{2} + .4 + 1$$

$$= 0.0007 + 0.0032 + 0.0213 + 0.08 + .4 + 1 = 1.505$$

$$y^{(4)} = \frac{(.4)^8}{384} + \frac{(.4)^7}{105} + \frac{(.4)^6}{48} + \frac{(.4)^5}{15} + \frac{(.4)^4}{8} + \frac{(.4)^3}{3} + \frac{(.4)^2}{2} + .4 + 1 = 1.505$$

(f) $x = 0.5$

$$y^{(1)} = \frac{(.5)^2}{2} + 0.5 + 1 = 0.125 + 0.5 + 1 = 1.625$$

$$y^{(2)} = \frac{(.5)^4}{8} + \frac{(.5)^3}{3} + \frac{(.5)^2}{2} + 0.5 + 1$$

$$= 0.0078 + 0.0417 + 0.125 + 0.5 + 1 = 1.675$$

$$y^{(3)} = \frac{(.5)^6}{48} + \frac{(.5)^5}{15} + \frac{(.5)^4}{8} + \frac{(.5)^3}{3} + \frac{(.5)^2}{2} + .5 + 1$$

$$= 0.0003 + 0.0021 + 0.0078 + 0.0417 + 0.125 + 0.5 + 1 = 1.677$$

$$y^{(4)} = \frac{(.5)^8}{384} + \frac{(.5)^7}{105} + \frac{(.5)^6}{48} + \frac{(.5)^5}{15} + \frac{(.5)^4}{8} + \frac{(.5)^3}{3} + \frac{(.5)^2}{2} + .5 + 1 = 1.677$$

5.19 THE RUNGE-KUTTA METHOD

In this method, the increments of the function (of functions) are calculated once for all by using a definite set of formulae. it means that necessary calculations for any increment are the same.

Let $$\frac{dy}{dx} = f(x, y)\ ;\ h = \text{interval}$$

x_0, y_0 are the initial values of x and y.

The following are the fourth order Runge-Kutta formulae which are most commonly used in practice.

$$k_1 = hf(x_0, y_0)$$

$$k_2 = hf\left(x_0 + \frac{h}{2}, y_0 + \frac{k_1}{2}\right)$$

$$k_3 = hf\left(x_0 + \frac{h}{2}, y_0 + \frac{k_2}{2}\right) \quad \text{...(A)}$$

$$k_4 = hf(x_0 + h, y_0 + k_3)$$

$$\Delta y = \frac{1}{6}[k_1 + 2(k_2 + k_3) + k_4]$$

$$\therefore \quad x_1 = x_0 + h \text{ and } y_1 = y_0 + \Delta y$$

For computing the increment in y for the second interval, the above formula may be used by simply changing $x_0, \to x_1, y_0 \to y_1$, *i.e.*,

$$k_1 = hf(x_1, y_1)$$

$$k_2 = hf\left(x_1 + \frac{h}{2}, y_1 + \frac{k_1}{2}\right)$$

$$k_3 = hf\left(x_1 + \frac{h}{2}, y_1 + \frac{k_2}{2}\right)$$

$$k_4 = hf(x_1 + h, y_1 + k_3)$$

$$\Delta y = \frac{1}{6}[k_1 + 2(k_2 + k_3) + k_4]$$

The process may be continued for evaluating the increment in y for the 3rd, 4th, 5th, ..., nth interval.

The method is explained with the help of the following solved examples.

Example 5.42. *Applying the Runge-Kutta method, find the solution of the equation for one step only* $\frac{dy}{dx} = xy^{1/3}$, *initial values are :* $x_0 = 1$, $y(0) = 1$ *and the interval* $h = 0.1$.

Solution:

$$\frac{dy}{dx} = f(x, y) = xy^{1/3}$$

$$x_0 = 1; y_0 = 1; h = 0.1$$

$$k_1 = hf(x_0, y_0)$$

$$= 0.1(1 \times 1^{1/3}) = 0.1$$

$$k_2 = hf\left(x_0 + \frac{h}{2}, y_0 = \frac{k_1}{2}\right)$$

$$= 0.1(1 + 0.05)\ (1 + 0.05)$$

$$= 0.1 \times 1.05 \times 1.05^{1/3} \qquad (\because \quad x = 1.05;\ y = 1.05)$$

$$= 0.1\ (1.05)^{4/3}$$

$$= 0.107$$

$$k_3 = hf\left(x_0 + \frac{h}{2}, y_0 = \frac{k_2}{2}\right)$$

$$= 0.1(1.05)\ (1 + 0.0535)^{1/3}$$

$$= 0.1 \times 1.05 \times (1.0.0535)^{1/3}$$

$$= 0.1068$$

$$k_4 = hf(x_0 + h, y_0 + k_3)$$

$$= 0.1(1 + 0.1)\ (1 + 0.1068)^{1/3}$$

$$= 0.1 \times 1.1(1.1068)^{1/3}$$

$$= 0.114$$

$$\Delta y = \frac{1}{6}[k_1 + 2(k_2 + k_3) + k_4]$$

$$= \frac{1}{6}[0.1 + 2(0.107 + 0.1068) + 0.114]$$

$$= 0.107$$

$$\therefore \qquad y_{1.1} = y_0 + 0.107$$

$$= 1 + 0.107 = 0.107 \quad \textbf{Ans.}$$

Example 5.43. *Use Runge-Kutta method to obtain the numerical solution of* $\frac{dy}{dx} = x^2 + y^2$, *$y(1) = 1.5$ in the interval (1, 1.3) with $h = 0.1$.*

Solution:

$$\frac{dy}{dx} = x^2 + y^2$$

$$x_0 = 1\ y_0 = 1.5;\ h = 0.1$$

(*i*)

$$x = 1.1$$

$$k_1 = hf(x_0, y_0)$$

$$= 0.1(1^2 + 1.5^2) = 0.325$$

$$k_2 = hf\left(x_0 + \frac{h}{2}, y_0 + \frac{k_1}{2}\right)$$

$$= 0.1\ [(1 + 0.05)^2 + (1.5 + 0.163)^2]$$

$$= 0.1[(1.05)^2 + (1.663)^2]$$

$$= 0.3868$$

$$k_3 = hf\left(x_0 + \frac{h}{2}, y_0 + \frac{k_2}{2}\right)$$

$$= 0.1[(1 + 0.05)^2 + (1.5 + 0.1934)^2]$$

$$= 0.1[(1.05)^2 + 1.6934)^2]$$
$$= 0.512$$
$$k_4 = hf(x_0 + h, y_0 + k_3)$$
$$= 0.1[(1 + 0.1)^2 + (1.5 + 0.512)^2]$$
$$= 0.1[(1.1)^2 + (2.012)^2]$$
$$= 0.526$$
$$\Delta y = \frac{1}{6}[k_1 + 2(k_2 + k_3 + k_4]$$
$$= \frac{1}{6}[0.325 + 2(0.3868 + 0.512) + 0.526]$$
$$= 0.4414$$
$$\therefore \quad y_{1.1} = y_0 + \Delta y$$
$$= 1.5 + 0.4414$$
$$= 1.94$$

(*ii*) $x = 1.2$

$$k_1 = hf(x_1\ y_1)$$
$$= 0.1\ [(1.1)^2 + (1.94)^2] = 0.4974$$
$$k_2 = hf\left(x_1 + \frac{h}{2}, y_1 + \frac{k_1}{2}\right)$$
$$= 0.1[(1.1 + 0.05)^2 + (1.94 + 0.2487)^2]$$
$$= 0.1[(1.15)^2 + (2.1887)^2]$$
$$= 0.6113$$
$$k_3 = hf\left(x_1 + \frac{h}{2}, y_1 + \frac{k_2}{2}\right)$$
$$= 0.1[(1.1 + 0.05)^2 + (1.94 + 0.3056)^2]$$
$$= 0.1[(1.15)^2 + (2.24056)^2]$$
$$= 0.634$$
$$k_4 = hf(x_1 + h, y_1 + k_3)$$
$$= 0.1[(1.1 + 0.1)^2 + (1.94 + 0.0634)^2]$$
$$= 0.1[(1.2)^2 + (2.574)^2]$$
$$= 0.806$$
$$\Delta y = \frac{1}{6}(k_1 + 2(k_2 + k_3) + k_4)$$
$$= \frac{1}{6}(0.4974 + 2(0.6113 + 0.634) + 0.806)$$

$$= \frac{1}{6}(0.4974 + 2.4906 + 0.806)$$

$$= 0.64$$

$$\therefore \quad y_{1.2} = y_1 + \Delta y = 1.94 + 0.64$$

$$= 2.58$$

$$x = 1.3$$

$$k_1 = hf(x_2, y_2)$$

$$= 0.1[(1.2)^2 + (2.58)^2]$$

$$= 0.810$$

$$k_2 = hf\left(x_2 + \frac{h}{2}, y_2 + \frac{k_1}{2}\right)$$

$$= 0.1[(1.2 + 0.05)^2 + (2.58 + 0.405)^2]$$

$$= 0.1[1.25)^2 + (2.985)^2]$$

$$= 1.057$$

$$k_3 = hf\left(x_2 + \frac{h}{2}, y_2 + \frac{k_2}{2}\right)$$

$$= 0.1[(1.2 + 0.05)^2 + (2.58 + 0.524)^2]$$

$$= 0.1[(1.25)^2 + (3.104)^2]$$

$$= 1.120$$

$$k_4 = hf(x_2 + h, y_2 + k_3)$$

$$= 0.1[(1.2 + 0.1)^2 + (2.58 + 1.120)^2]$$

$$= 0.1[(1.2)^2 + (3.70)^2]$$

$$= 1.538$$

$$\Delta y = \frac{1}{6}[k_1 + 2(k_2 + k_3) + k_4]$$

$$= \frac{1}{6}(0.81 + 2(1.047 + 1.120) + 1.538)$$

$$= 1.114$$

$$y_{1.3} = y_2 + \Delta y$$

$$= 2.58 + 1.114 = 3.694$$

$$\left.\begin{array}{ll} x = 1 & y = 1.5 \\ x = 1.1 & y = 1.94 \\ x = 1.2 & y = 2.58 \\ x = 1.3 & y = 3.694 \end{array}\right] \textbf{Ans.}$$

Example 5.44. *Show one step of the fourth order Runge-Kutta method for the solution of the following differential equation*

$$\frac{dy}{dx} = x + y^2 \; ; y(0) = 1.0$$

Find y at x = 0.5

Solution: Here, $f(x, y) = x + y^2$; $x_0 = 0$ $y_0 = 1$ and $h = 0.5$

$$k_1 = hf(x_0, y_0)$$
$$= 0.5(0 + 1^2) = 0.05000$$
$$k_2 = hf\left(x_0 + \frac{h}{2}, y_0 + \frac{k_1}{2}\right)$$
$$= 0.5[0.25 + (1 + 0.250]^2$$
$$= 0.5[0.25 + (1.25)^2]$$
$$= 0.906$$
$$k_3 = hf\left(x_0 + \frac{h}{2}, y_0 + \frac{k_2}{2}\right)$$
$$= 0.5[0.25 + (1 + 0.453)^2]$$
$$= 1.180$$
$$k_4 = hf(x_0 + h + y_0 + k_3)$$
$$= 0.5[0.5 + (1 + 1.180)^2]$$
$$= 2.625$$
$$\Delta y = \frac{1}{6}(k_1 + 2(k_2 + k_3) + k_4)$$
$$= \frac{1}{6}[0.5 + 2(0.906 + 1.180) + 2.626]$$
$$= 1.216$$

$\therefore$ $$y(0.5) = y_0 + \Delta y$$
$$= 1 + 1.216 = 2.216$$ **Ans.**

Alternate method:

Given equation: $\frac{dy}{dx} = x + y = f(x, y)$

$$x_0 = 0, y_0 = 1.0 \quad \text{Let } h = 0.1$$

1st step

$$k_1 = hf(x_0, y_0) = 0.1\, f(0, 1)$$
$$= 0.1 \times (0 + 1^2) = 0.1$$
$$k_2 = hf\left(x_0 + \frac{h}{2}, y_0 + \frac{k_1}{1}\right)$$

$$= 0.1f(0 + 0.05, 1 + 0.05)$$
$$k_2 = 0.1 \times (0.05 + 1.05^2) = 0.11525$$
$$k_3 = hf\left(x_0 + \frac{h}{2}, y_0 + \frac{k_2}{2}\right)$$
$$= 0.1f(0 + 0.05, 1 + 0.057625)$$
$$k_3 = 0.1 \times (0.05 + 1.57625^2) = 0.116857$$
$$k_4 = hf(x_0 + h, y_0 + k_3)$$
$$= 0.1\, f(0 + 0.1, 1 + 0.116857)$$
$$= 0.1 \times f(0.1, 1.116857)$$
$$= 0.1 \times (0.1 + 1.116857^2) = 0.1347369$$
$$y_{(0.1)} = y_0 + \frac{1}{6}(k_1 + 2k_2 + 2k_3 + k_4)$$
$$= 1 + \frac{1}{6}(0.1 \times 2 \times 0.11525 \times 0.116857 + 0.1347369)$$
$$= 1 + 0.1164918 = 1.1164918$$

Proceed further till a desired result is obtained.

Examples. 5.45. *Solve the following differential equation using the 4th order Runge-Kutta method*

$$\frac{dy}{dx} = 0.5\,(1 + x)y^2$$
$$x_0 = 1 \text{ and } y_0 = 1$$

Take a step size h = 0.1 and solve upto x = 0.3

Solution:

$$\frac{dy}{dx} = 0.5\,(1 + x)y^2$$
$$x_0 = 1 \text{ and } y_0 = 1, h = 0.1$$

(1) $$k_1 = hf(x_0, y_0)$$
$$= 0.1 \times 0.5\,(1 + 1)1^2$$
$$= 0.1 \times 1 = 0.1$$
$$k_2 = hf\left(x_0 + \frac{h}{2}, y_0 + \frac{k_1}{2}\right)$$
$$(1 + 0.05)\,(1 + 0.05)$$
$$= 0.1[0.5\,(1 + 1.05)\,(1.05)^2]$$
$$= 0.1[0.5 \times 2.05 \times (1.05)^2]$$
$$= 0.113$$

$$
\begin{aligned}
k_3 &= h f\left(x_0+\frac{h}{2}, y_0+\frac{k_2}{2}\right)\\
&= 0.1\ [0.5\ (1+1.05)\ (1+0.0565)^2]\\
&= 0.1 \times 0.5 \times 2.05 \times (1.0565)^2\\
&= 0.114\\
k_4 &= hf(x_0+h, y_0+k_3)\\
&= 0.1\ [0.5\ (1+1.1)\ (1+0.114)^2]\\
&= 0.1 \times 0.5 \times 2.1 \times (1.114)^2\\
&= 0.130\\
\Delta y &= \frac{1}{6}[k_1+2\ (k_2+k_3)+k_4]\\
&= \frac{1}{6}[0.1+2\ (0.113+0.114)+0.130]\\
&= 0.114
\end{aligned}
$$

$$
\begin{aligned}
\therefore \quad y_1 &\doteq y_0+\Delta y\\
&= 1+0.114 = 1.114\\
x_1 &= x_0+h = 1+0.1 = 1.1;\ y_1 = 1.114\\
k_1 &= hf(x_1, y_1)\\
&= 0.1[0.5\ (1+1.1)\ (1.114)^2]\\
&= 0.1 \times 0.5 \times 2.1 \times 1.114^2\\
&= 0.130\\
k_2 &= h f\left(x_1+\frac{h}{2}, y_1+\frac{k_1}{2}\right)\\
&= 1.1+0.05,\ 1.114+0.065\\
&= 0.1[0.5\ (1+1.15)\ (1.179)^2]\\
&= 0.149\\
k_3 &= h f\left(x_1+\frac{h}{2}, y_1+\frac{k_2}{2}\right)\\
&= (1.1+0.05)\ (1.114+0.0745)\\
&= 0.1\ [0.5\ (2.15)\ (1.1885)^2]\\
&= 0.1518\\
k_4 &= hf(x_1+h, y_1+k_3)\\
&= 1.1+0.1,\ \ 1.114+0.1518\\
&= 0.1\ [0.5\ (1+1.2)\ (1.2658)^2]\\
&= 0.1\ [(0.5 \times 2.2) \times (1.2658)^2]\\
&= 0.176
\end{aligned}
$$

$$\Delta = \frac{1}{6}[k_1 + 2(k_2 + k_3) + k_4]$$

$$= \frac{1}{6}[0.13 + 2(0.149 + 0.1518) + 0.176]$$

$$= 0.151$$

$$\therefore \quad y_2 = y_1 + \Delta y$$

$$= 1.114 + 0.151 = 1.265$$

$$x_2 = x_1 + h = 1.1 + 0.1 = 1.2 \; ; y_2 = 1.265$$

$$k_1 = h f(x_2, y_2)$$

$$= 0.1 [0.5 (1 + 1.2) (1.265)^2]$$

$$= 0.176$$

$$k_2 = h f\left(x_2 + \frac{h}{2}, y_2 + \frac{k_1}{2}\right)$$

$$= 0.1 [0.5 (1 + 1.2 + 0.05) (1.265 + 0.088))^2]$$

$$= 0.1 [0.5 (2.25) (1.353)^2]$$

$$= 0.206$$

$$k_3 = hf\left(x_2 + \frac{h}{2}, y_2 + \frac{k_2}{2}\right)$$

$$= 0.1[0.5 (1 + 1.2 + 0.05) (1.265 + 0.103)^2]$$

$$= 0.1[0.5 (2.25) (1.368)^2]$$

$$= 0.210$$

$$k_4 = hf(x_2 + h, y + k_3)$$

$$= 0.1[0.5 (1 + 1.2 + 0.1) (1.265 + 0.210)^2]$$

$$0.1[0.5 \times 2.3 \times (1.4757)^2]$$

$$= 0.250$$

$$\Delta y = \frac{1}{6}[k_1 + 2(k_2 + k_3) + k_4]$$

$$= \frac{1}{6}[0.176 + 2(0.206 + 0.210) + 0.250]$$

$$= 0.210$$

$$\therefore \quad y_3 = y_2 + \Delta y$$

$$= 1.265 + 0.210 = 1.475$$

$$x_0 = 1, y_0 = 1$$

$$\left.\begin{array}{ll} x_{(.1)} = 1.1, & y_{(.1)} = 1.114 \\ x_{(.2)} = 1.2, & y_{(.2)} = 1.265 \\ x_{(.3)} = 1.3, & y_{(.3)} = 1.475 \end{array}\right\} \textbf{Ans.}$$

Example. 5.46. *By Range-Kutta method of fourth order, solve the following equation*

$$\frac{dy}{dx} = \frac{y}{x}, y(1) = 1.$$

Obtain when x = 1.2 in steps of 0.1.

Solution: Given:

$$x_0 = 1, y_0 = 1, h = 0.1$$

$$k_1 = hf(x_0, y_0)$$

$$= 0.1\frac{1}{1} = 0.1$$

$$k_2 = hf\left(x_0 + \frac{h}{2}, y_0 + \frac{k_1}{2}\right)$$

$$= 0.1[1 + 0.05, 1 + 0.05]$$

$$= 0.1\left(\frac{1.05}{1.05}\right) = 0.1$$

$$k_3 - hf\left(x_0 + \frac{h}{2}, y_0 + \frac{k_1}{2}\right)$$

$$= 0.1(1 + 0.05, 1 + 0.05)$$

$$= 0.1\left(\frac{1.05}{1.05}\right) = 0.1$$

$$k_4 = hf(x_0 + h, y_0 + k_3)$$

$$= 0.1(1 + 0.1, 1 + 0.1)$$

$$= 0.1\left(\frac{1.1}{1.1}\right) = 0.1$$

$$\Delta y = \frac{1}{6}[k_1 + 2(k_2 + k_3) + k_4]$$

$$= \frac{1}{6}(0.1 + 2(.1 + .1) + .1)$$

$$= \frac{.6}{6} = .1$$

$$\therefore \quad y_1 = y_0 + .1 = 1 + .1 = 1.1$$

$$x_1 = x_0 + h = 1 + .1 = 1.1 ; y_0 = 1.1$$

$$k_1 = hf(x_1, y_1)$$

$$= 0.1(1.1/1.1) = 0.1$$

$$k_2 = \left(x_1 + \frac{h}{2}, y_1 + \frac{k_1}{2}\right)$$
$$= 0.1\ (1.1 + 0.05,\ 1.1 + 0.05$$
$$= 0.1\left(\frac{1.15}{1.15}\right) = 0.1$$

$$k_3 = hf\left(x_1 + \frac{h}{2}, y_1 + \frac{k_2}{2}\right)$$
$$= 0.1\ (1.1 + 0.05,\ 1.1 + 0.05)$$
$$= 0.1\left(\frac{1.15}{1.15}\right) = 0.1$$

$$k_4 = hf(x_1 + h, y_1 + k_3)$$
$$= 0.1\ (1.1 + 0.1,\ 1 + 0.1)$$
$$= 0.1\left(\frac{1.2}{1.2}\right) = 0.1$$

$$\Delta y = \frac{1}{6}[k_1 + 2\ (k_2 + k_3) + k_4]$$
$$= \frac{1}{6}(.1 + 2\ (.1 + .1) + .1)$$
$$= 0.1$$
$$y_2 = y_1 + dy$$
$$= 1.1 + .1 = 1.2$$
$$\therefore \quad y_2 = 1.2 \text{ when } x = 1.2 \textbf{ Ans.}$$

Example. 5.47. *Solve the equation by Runge-Kutta method*

$$\frac{dy}{dx} = \frac{y - x}{y + x}$$

for x = 0.1, 0.2, 0.3 with initial conditions $x_0 = 0$, $y_0 = 1$

Solution: Given equation is

$$\frac{dy}{dx} = \frac{y - x}{y + x} = f(x, y)$$

$$\therefore \quad f(x, y) = \frac{y - x}{y + x},\ x_0 = 0,\ y_0 = 1$$

Let $h = 0.1$

$$\therefore \quad k_1 = hf(x_0, y_0) = 0.1\, f(0, 1) = 0.1 \times \frac{1-0}{1+0} = 0.1$$

$$k_2 = hf\left(x_0 + \frac{h}{2}, y_0 + \frac{k_1}{2}\right) = 0.1\, f(0 + 0.05, 1 + 0.05)$$

$$= 0.1f(0.05, 1.05)$$

$$k_2 = 0.1 \times \frac{1.05 - 0.05}{1.05 + 0.05} = 0.1 \times \frac{1.0}{1.1} = \frac{1}{11} = 0.0909$$

$$k_3 = \left(x_0 + \frac{h}{2}, y_0 + \frac{k_2}{2}\right) = 0.1\, f(0 + 0.05, 1 + 0.04545)$$

$$= 0.1\, f(0.05, 1.04545)$$

$$k_3 = 0.1 \times \frac{1.04545 - 0.05}{1.04545 + 0.05} = 0.1 \times \frac{0.99545}{1.099545} = 0.0909$$

$$k_4 = hf(x_0 + h, y_0 + k_3) = 0.1\, f(0.05, 1 + 0.0909)$$

$$= 0.1\, f(0.05, 1.0909)$$

$$= 0.1\, \frac{1.0909 - 0.05}{1.0909 + 0.05} = 0.1 \times \frac{1.0409}{1.1409} = 0.0912$$

$$y_{(0.1)} = y_0 + \frac{1}{6}(k_1 + 2k_2 + 2k_3 + k_4)$$

$$= 1 + \frac{1}{6}(0.1 + 2 \times 0.0909 + 2 \times 0.0909 + 0.0912)$$

$$= 1 + 0.0925$$

$$= 1.0925$$

$$\therefore \quad y_{(0.1)} = 1.0925$$

Now, taking $x_0 = 0.1$, $y_0 = 1.0925$, $h = 0.1$, repeat the above procedure.

$k_1 = h\, f(x_0, y_0) = 0.1\, f(0.1, 1.0925)$

$$= 0.1 \times \frac{1.0925 - 0.1}{1.0925 + 0.1} = 0.1 \times \frac{0.9925}{1.1925} = 0.0834$$

$$k_2 = h\, f\left(x_0 + \frac{h}{2}, y_0 + \frac{k_1}{2}\right)$$

$$= 0.1\, f(0.1 + 0.05, 1.0925 + 0.0417)$$

$$= 0.1\, f(0.15, 1.1342)$$

$$k_2 = 0.1 \times \frac{1.1342 - 0.15}{1.342 + 0.15} = 0.1 \times \frac{0.9842}{1.2842} = 0.0766$$

$$k_3 = \left(x_0 + \frac{h}{2}, y_0 + \frac{k_2}{2}\right)$$
$$= 0.1\, f\,(0.15,\ 1.0925 + 0.0383)$$
$$= 0.1\, f\,(0.15,\ 1.1308)$$
$$= 0.1 \times \frac{1.1308 - 0.15}{1.1308 + 0.15} = 0.1 \times \frac{0.9808}{1.2808} = 0.0766$$
$$k_4 = h\, f\,(x_0 + h,\ y_0 + k_3)$$
$$= 0.1 f(0.1 + 0.1, 1.0925 + 0.0766) = 0.1 f(0.2, 1.1691)$$
$$= 0.1 \times \frac{1.1691 - 0.2}{1.1691 + 0.2} = 0.1 \times \frac{0.9691}{1.3691} = 0.0708$$

$$\therefore \quad y_{(0.2)} = y_0 + \frac{1}{6}(k_1 + 2k_2 + 2k_3 + k_4)$$
$$= 1.0925 + \frac{1}{6}(0.0834 + 2 \times 0.0766 + 2 \times 0.0766 + 0.0708)$$
$$= 1.0925 + \frac{1}{6}(0.0834 + 0.1532 + 0.1532 + 0.0708)$$
$$= 1.0925 + \frac{1}{6}(4606)$$
$$= 1.0933 + 0.0768$$
$$= 1.1701$$

To find $y_{(0.3)}$, start the calculations with $x_0 = 0.2$, $y_0 = 1.1701$ and proceed as above. **Ans.**

Example. 5.48. *Solve the following differential equation using the 4th order Runge-Kutta method*

$$\frac{dy}{dx} = 0.5\,(1 + x)y^2$$
$$x_0 = 1 \text{ and } y_0 = 1$$

Solution: 1st Step:

$$x_0 = 0;\ y_0 = 1,\ h = 0.1$$
$$k_1 = h\, f\,(x_0, y_0) = 0.1\, f\,(0,1)$$
$$= 0.1 \times 0.5\,(1 + 0) \times 1 = 0.05$$
$$k_2 = h f\left(x_0 + \frac{h}{2}, y_0 + \frac{k_1}{2}\right) = 0.1\, f\,(0 + 0.05,\ 1.025)$$
$$= 0.1 \times 0.5\,(1 + 0.05) \times (1.025)^2 = 0.05516$$

$$k_3 = hf\left(x_0 + \frac{h}{2}, y_0 + \frac{k_2}{2}\right) = 0.1\,f(0 + 0.05, 1.02758)$$

$$= 0.1 \times 0.5\,(1 + 0.05) \times (1.02758)^2 = 0.05544$$

$$k_4 = hf(x_0 + h, y_0 + k_3) = 0.1\,f(0 + 0.1, 1 + 05544)$$

$$= 0.1 \times 0.5 \times (1 + 01) \times (1.05544)^2 = 0.061267$$

$$y(0.1) = y_0 + \frac{1}{6}[k_1 + 2k_2 + 2k_3 + k_4]$$

$$= 1 + \frac{1}{6}(0.05 + 2 \times 0.05516 + 2 \times 0.05544 + 0.061267)$$

$$= 1 + \frac{1}{6}(0.05 + 0.11032 + 0.11088 + 0.061267)$$

$$= 1.055411$$

2nd Step:

$$x_0 = 0.1, \quad y_0 = 1.055411, \quad h = 0.1$$

$$k_1 = hf(x_0, y_0) = 0.1\,f(0.1, 1.055411)$$

$$= 0.1 \times 0.5 \times (1 + 0.1) \times (1.055411)^2 = 0.061264$$

$$k_2 = hf\left(x_0 + \frac{h}{2}, y_0 + \frac{k_1}{2}\right)$$

$$= 0.1\,f(0 + 0.05, 1.055411 + 0.030632)$$

$$= 0.1\,f(0.15, 1.086043)$$

$$= 0.1 \times 0.5\,(1 + 0.15) \times (1.086043)^2 = 0.0678206$$

$$k_3 = hf\left(x_0 + \frac{h}{2}, y_0 + \frac{k_2}{2}\right)$$

$$= 0.1\,f(0.1 + 0.05, 1.055411 + 0.0339103)$$

$$= 0.1\,f(0.15, 1.086043)$$

$$= 0.1 \times 0.5\,(1 + 0.15 \times (1.086043)^2 = 0.0678206$$

$$k_4 = hf(x_0 + h, y_0 + k_3)$$

$$= 0.1\,f(0.1 + 0.1, 1.055411 + 0.0682306)$$

$$= 0.1\,f(0.2, 1.1236416)$$

$$= 0.1 \times 0.5 \times (1.2) \times (1.1236416)^2 = 0.0757542$$

$$y(0.2) = y_0 + \frac{1}{6}(k_1 + 2k_2 + 2k_3 + k_4)$$

$$= 1.055411 + \frac{1}{6}(0.061264 + 2 \times 0.0678206 + 2 \times 0.0682306 + 0.0757542)$$

$$= 1.055411 + \frac{1}{6}(0.061264 + 0.1356412 + 0.1364612 + 0.0757542)$$

i.e., $y(0.2) = 1.055411 + 0.0681867 = 1.1235977$

3rd Step:

$$x_0 = 0.2, \qquad y_0 = 1.235977, \qquad h = 0.1$$

$$k_1 = h\,f(x_0, y_0) = 0.1\,f(0.2, 1.1235977)$$

$$= 0.1 \times 0.5 \times (1 + 0.2)\,(1.1235977)^2 = 0.0757483$$

$$k_2 = h\,f\left(x_0 + \frac{h}{2}, y_0 + \frac{k_1}{2}\right)$$

$$= 0.1\,f(0.2 + 0.05, 1.1235977 + 0.037842)$$

$$= 0.1\,f(0.25, 1.1614719)$$

$$= 0.1 \times 0.5\,(1 + 0.25)\,(1.1614719)^2 = 0.0843135$$

$$k_3 = h\,f\left(x_0 + \frac{h}{2}, y_0 + \frac{k_2}{2}\right)$$

$$= 0.1\,f(0.2 + 0.05, 1.1235977 + 0.0421568)$$

$$= 0.1\,f(0.25, 1.1657545)$$

$$= 0.1 \times 0.5\,(1 + 0.25 \times (1.165745)^2 = 0.0849364$$

$$k_4 = h\,f(x_0 + h, y_0 + k_3)$$

$$= 0.1\,f(0.2 + 0.1, 1.1235977 + 0.0849364)$$

$$= 0.1\,f(0.3, 1.2085341)$$

$$= 0.1 \times 0.5 \times (1 + 0.3) \times (1.2085341)^2 = 0.094936$$

$$y_{(0.3)} = y_0 + \frac{1}{6}(k_1 + 2k_2 + 2k_3 + k_4)$$

$$= 1.1235977 + \frac{1}{6}(0.07577483 + 2 \times 0.0843135 + 2 \times 0.0849364 + 0.094936)$$

$$= 1.1235977 + \frac{1}{6}(0.075483 + 2 \times 0.168627 + 0.1698728 + 0.094936)$$

$$= 1.1235977 + 0.084864$$

$$y_{10.3} = 1.2084617$$

$\therefore$ When $x = 0.3, y = 1.2084617$. **Ans.**

5.20 SOLUTION OF DIFFERENTIAL EQUATIONS BY RANGE-KUTTA METHOD

Differential equations of higher order may be solved by Range-Kutta method by reducing them to the first order simultaneous differential equations.

(*i*) **Simultaneous Equations.** Let the given differential equations be

$$\frac{dy}{dx} = f_1(x, y, z)$$

$$\frac{dy}{dx} = f_2(x, y, z)$$

with the initial conditions $y = y_0$, $z = z_1$ when $x = x_0$ and increment $= h$.

To find the values of y and z say y_1 and z_1 when $x = x_0 + h = x_1$ proceed as under :

$y_1 = y_0 + k$ and $z_1 = z_0 + 1$

$$k_1 = h\,f_1(x_0, y_0, z_0), \qquad l_1 = h\,f_2\,(x_0, y_0, z_0)$$

$$k_2 = hf_1(x_0 + \frac{h}{2}, y_0 + \frac{k_1}{2}, z_0 + \frac{l_1}{2}) \qquad l_2 = hf_2(x_0 + \frac{h}{2}, y_0 + \frac{k_1}{2}, z_0 + \frac{l_1}{2})$$

$$k_3 = hf_1(x_0 + \frac{h}{2}, y_0 + \frac{k_2}{2}, z_0 + \frac{l_2}{2}); \qquad l_3 = hf_2(x_0 + \frac{h}{2}, y_0 + \frac{k_2}{2}, z_0 + \frac{l_2}{2})$$

$$k_4 = hf_1(x_0 + h, y_0 + k_3, z_0 + l_3); \qquad l_4 = hf_2(x_0 + h, y_0 + k_3, z_0 + l_3)$$

$$k = \frac{1}{6}(k_1 + 2(k_2 + k_3) + k_4); \qquad l = \frac{1}{6}(l_1 + 2(l_2 + l_3) + l_4)$$

To obtain the values of y and z say y_2 and z_2, for the next increment for x from x_1 to x_2 where $x_2 = x_1 + h$, replace x_0, y_0, z_0 by x_1, y_1, z_1 respectively. From the above results the method may be applied for obtaining the further values y_3, z_3 y_4, z_4 y_5, z_5, etc.

Examples. 5.49. *Solve the equations*

$$\frac{dy}{dx} = xz + 2\,; \quad \frac{dz}{dx} = -\,xy$$

for x = 0.2 (0.2) 0.6.

Given: $\quad y = 0, z = 1$ *when* $x = 0$

Solution: $f_1(x, y, z) = zx + 2;\;\; f_2\,(x, y, z) = -\,xy$

and $\quad x_0 = 0, y_0 = 0, z_0 = 1$

and $\quad h = 0.2$

When $x = 0.2$

$$\begin{aligned} k_1 &= hf_1(x_0, y_0, z_0) \\ &= 0.2(0, 0, 1) \\ &= 0.2(0 \times 1 + 2) = 0.2 \times 2 = 0.4 \\ l_1 &= hf_2(x_0, y_0, z_0) \\ &= 0.2(0, 0, 1) \\ &= 0.2(-\,0 \times 0) = 0.0 \end{aligned}$$

$$k_2 = hf_1\left(x_0 + \frac{h}{2}, y_0 + \frac{k_1}{2}, z_0 + \frac{l_1}{2}\right)$$
$$= 0.2f_1(0.1, 0.2, 1 + 0)$$
$$= 0.2(0.1 \times 1 + 2) = 0.42$$

$$l_2 = hf_2\left(x_0 + \frac{h}{2}, y_0 + \frac{k_1}{2}, z_0 + \frac{l_1}{2}\right)$$
$$= 0.2\, f_2(0.1, 0.2, 1)$$
$$= 0.2(-\,0.1 \times 0.2) = -\,0.004$$

$$k_3 = hf_1\left(x_0 + \frac{h}{2}, y_0 + \frac{k_2}{2}, z_0 + \frac{l_2}{2}\right)$$
$$= 0.2\, f_1(0.1, 0.21, 0.998)$$
$$= 0.2(0.1 \times 0.21 + 2) = -\,0.4042$$

$$l_3 = hf_1\left(x_0 + \frac{h}{2}, y_0 + \frac{k_2}{2}, z_0 + \frac{l_2}{2}\right)$$
$$= 0.2\, f_2(0.1, 0.21, 0.998)$$
$$= 0.2\,(-\,0.1 \times 0.21) = -\,0.0042$$
$$k_4 = hf_1(x_0 + h, y_0 + k_3, z_0 + l_3)$$
$$= 0.2\, f(0.2, 0.4042, 0.9958)$$
$$= 0.2(0.2 \times 0.9958 + 2) = 0.4398$$
$$l_4 = hf_2(x_0 + h, y_0 + k_3, z_0 + l_3)$$
$$= 0.2\, f_2(0.2, .4042, 0.9958)$$
$$= 0.2(-\,0.2 \times 0.4042) = -\,0.016$$

$$k = \frac{1}{6}\,[k_1 + 2(k_2 + k_3) + k_4]$$
$$= \frac{1}{6}\,(0.4 + 2(0.42 + 0.4042) + 0.4398)$$
$$= 0.415$$

$$l = \frac{1}{6}\,(l_1 + 2(l_2 + l_3) + l_4)$$
$$= \frac{1}{6}\,(0.0 + 2(-\,0.004 - 0.0042) - 0.016)$$
$$= \frac{1}{6}\,(-\,0.0324) = -\,0.0054$$

$$y_1 = y_0 + k = 0 + 0.410 = 0.415$$
$$z_1 = z_0 + 1 = 1 + (-0.0054) = 0.9946$$

When $x = 0.4$

$$x_1 = 0.2\ ;\ y_1 - 0.415,\ z_1 = 0.995$$
$$\begin{aligned} k_1 &= hf_1(x_1, y_1, z_1) \\ &= hf_1(xz + 2) \\ &= 0.2f_1(0.2 \times 0.995 + 2) \\ &= 0.2(2.082) = 0.4398 \end{aligned}$$
$$\begin{aligned} l_1 &= hf_2(x_1, y_1, z_1) \\ &= hf(-x_1y_1) \\ &= 0.2(-0.2 \times 0.415) \\ &= -0.0166 \end{aligned}$$
$$\begin{aligned} k_2 &= hf_1\left(x_1 + \frac{h}{2}, y_1 + \frac{k_1}{2}, z_1 + \frac{l_1}{2}\right) \\ &= h(0.2 + 0.1, 0.415 + 0.2199, 0.995 + (-0.083)) \\ &= 0.2\ f_1(0.3, 0.635, 0.912) \\ &= 0.2(0.3 \times 0.912 + 2) \\ &= 0.455 \end{aligned}$$
$$\begin{aligned} l_2 &= \left(x_2 + \frac{h}{2}, y_1 + \frac{k_1}{2}, z_1 + \frac{l_1}{2}\right) \\ &= 0.2\ f_2(0.2, 0.635, 0.913) \\ &= 0.2(-0.3 \times 0.535) = -0.038 \end{aligned}$$
$$\begin{aligned} k_3 &= hf_1\left(x_1 + \frac{h}{2}, y_1 + \frac{k_2}{2}, z_1 + \frac{l_2}{2}\right) \\ &= 0.2\ f_1(0.2 + 0.1, 0.415 + 0.228, 0.995 - 0.019) \\ &= 0.2\ f_1(0.3, 0.643, 0.976) \\ &= 0.2(0.3 \times 0.976 + 2) \\ &= 0.4586 \end{aligned}$$
$$\begin{aligned} l_3 &= hf_2\left(x_1 + \frac{h}{2}, y_1 + \frac{k_2}{2}, z_1 + \frac{l_2}{2}\right) \\ &= 0.2\ f_2(0.2 + 0.1, 0.415 + 0.228, 0.995 - 0.019) \\ &= 0.2(0.3, 0.643, 0.976) \\ &= 0.2(-0.3 \times 0.643) \\ &= -0.38 \end{aligned}$$
$$k_4 = hf_1(x_1 + h, y_1 + k_3, z_1 + l_3)$$

$$= 0.2\, f_2(0.2 + 0.2, 0.415 + 0.4586, 0.995 - 0.038)$$
$$= 0.2\, f_1(0.2, 0.874, 0.957)$$
$$= 0.2(0.4 \times 0.957 + 2) = 0.477$$
$$l_4 = hf_2(x_1 + h, y_1 + k_3, z_1 + l_3)$$
$$= 0.2\, f_1(0.4, 0.874, 0.957)$$
$$= 0.2(-0.4 \times 0.874) = -0.070$$
$$k = \frac{1}{6}[k_1 + 2(k_2 + k_3) + k_4]$$
$$= \frac{1}{6}[0.440 + 2(0.455 + 0.459) + 0.477] = 0.458$$
$$l = \frac{1}{6}[l_1 + 2(l_2 + l_3) + l_4]$$
$$= \frac{1}{6}[-0.017 + 2(-0.038 - 0.038) - 0.07] = -0.04$$
$$y_2 = y_1 + k = 0.415 + 0.458 = 0.873$$
$$z_2 = z_1 + l = 0.995 + (-0.040) = 0.955$$

When $x = 0.6$

$$x_2 = 0.4, y_2 = 0.873, z_2 = 0.955$$
$$k_1 = hf_1(x_2, y_2, z_2) = 0.2f_1(xz + 2)$$
$$= 0.2f_1(0.4 \times 0.995 + 2) = 0.476$$
$$l_1 = hf_2(x_2, y_2, z_2) = 0.2f_2(-xy)$$
$$= 0.2(-0.4 \times 0.8730) = -0.070$$
$$k_2 = hf_1\left(x_2 + \frac{h}{2}, y_2 + \frac{k_1}{2}, z_2 + \frac{l_1}{2}\right)$$
$$= 0.2f_1(0.4 + 0.1, 0.673 + 0.238, 0.955 + (-0.035)$$
$$= 0.2f_1(0.5, 1.111, 0.92)$$
$$= 0.2(0.5 + 0.92 + 2) = 0.492$$
$$l_2 = hf_2(0.4, 1.111, 0.92)$$
$$= 0.2(-0.5 \times 1.111) = -0.111$$
$$k_3 = hf_1\left(x_2 + \frac{h}{2}, y_2 + \frac{k_2}{2}, z_2 + \frac{l_2}{2}\right)$$
$$= 0.2(0.4 + 0.1, 0.873 + 0.246, 0.955 + (-0.056)$$
$$= 0.2(0.5, 1.119, 0.899)$$
$$= 0.2(0.5 \times 0.899 + 2) = 0.490$$
$$l_3 = hf_2\left(x_2 + \frac{h}{2}, y_2 + \frac{k_2}{2}, z_2 + \frac{l_2}{2}\right)$$

$$\begin{aligned}
&= 0.2f_2\,(0.5,\ 0.873 + 0.246,\ 0.955 - 0.056)\\
&= 0.2(0.5,\ 1.119,\ 0.899)\\
&= 0.2(-0.5 \times 1.119) = -0.112\\
k_4 &= hf_1(x_2 + h,\ y_2 + k_3,\ z_2 + l_3)\\
&= 0.2f_1\,(0.6,\ 0.873 + 0.49 + 0.955 - 0.056)\\
&= 0.2f_1\,(0.6,\ 1.363,\ 0.899)\\
&= 0.2(0.5 \times 0.899 + 2) = 0.508\\
l_4 &= hf_2(x_2 + h,\ y_2 + k_3,\ z_2 + l_3)\\
&= 0.2f_2\,(0.6,\ 1.363,\ 0.899)\\
&= 0.2(-0.6 \times 1.363) = -0.164\\
k &= \frac{1}{6}[k_1 + 2(k_2 + k_3) + k_4]\\
&= \frac{1}{6}[0.476 + 2(0.492 + 490) + 0.508]\\
&= 0.491\\
l &= \frac{1}{6}[l_1 + 2(l_2 + l_3) + l_4]\\
&= \frac{1}{6}[-0.07 + 2(-0.111 - 0.112) - 0.164]\\
&= -0.113
\end{aligned}$$

$$\therefore \quad y_3 = y_2 + k = 0.875 + 0.491 = 1.366$$

$$z_3 = z_2 + l = 0.955 - 0.113 = 0.842$$

$$\left.\begin{aligned}
&x_0 = 0;\ y_0 = 0;\ z_0 = 1\\
&x_1 = .2;\ y_1 = 0.415;\ z_1 = 0.955\\
&x_2 = .4;\ y_2 = 0.873;\ z_2 = 0.955\\
&x_3 = .6;\ y_3 = 1.366;\ z_3 = 0.842.
\end{aligned}\right\}\ \textbf{Ans.}$$

(*ii*) Second Order Equations of the General Type

$$y'' = f(x, y, y')$$

Let $y' = z$, then $y'' = z'$

The given equation thus reduces to the following equations

$$\frac{dy}{dx} = z \qquad \text{...}(i)$$

and

$$\frac{dz}{dx} = f(x, y, z) \qquad \text{...}(ii)$$

The solution of these equations may be done as two simultaneous equations discussed earlier.

Example 5.50. *Solve the equation*

$$y'' = xy' - y^2 \text{ for } x = 0.2$$

Given: $y = 1, y' = 0$ *when* $x = 0$.

Solution: Let $y' = z$, then $y'' = z'$

Hence, the given equation reduces to the form

$$\frac{dy}{dx} = z; \frac{dz}{dx} = xz - y^2$$

Further $x_0 = 0, y_0 = 1, z_0 = 0$ and $h = 0.2$

Also, $f_1(x, y, z) = z$

$$f_2(x, y, z) = xz - y^2$$

Now,

$$\begin{aligned}
k_1 &= hf_1(x_0, y_0, z_0)\\
&= h \cdot z_0\\
&= 0.2 \times 0 = 0\\
l_1 &= hf_2(x_0, y_0 - z_0)\\
&= h(xz - y^2)\\
&= 0.2(0 \times 0 - 1^2) = -0.2\\
k_2 &= hf_1\left(x_0 + \frac{h}{2}, y_0 + \frac{k_1}{2}, z_0 + \frac{l_1}{2}\right)\\
&= h(0 + 1.1, 1 + 0, 0 - 0.1)\\
&= h(0.1, 1, -0.1)\\
&= 0.2(z) = 0.2 \times (-0.1) = -0.02\\
l_2 &= hf_2\left(x_0 + \frac{h}{2}, y_0 + \frac{k_1}{2}, z_0 + \frac{l_1}{2}\right)\\
&= h(0 + 0.1, 1 + 0, 0 - 0.1)\\
&= h(0.1, 1, -0.1)\\
&= h(xz - y^2)\\
&= 0.2(0.1 \times (-0.1) - 1^2)\\
&= 0.2(-0.01 - 1) = -1.01 \times 0.2 = -0.202\\
k_3 &= hf_1\left(x_0 + \frac{h}{2}, y_0 + \frac{k_2}{2}, z_0 + \frac{l_2}{2}\right)\\
&= hf_1(0 + 0.1, 1 - 0.01, 0 - 0.101)\\
&= hf_1(0.1, 0.99, -0.101)\\
&= 0.2(z) = 0.2 \times (-0.101) = -0.0202\\
l_3 &= hf_2\left(x_0 + \frac{h}{2}, y_0 + \frac{k_2}{2}, z_0 + \frac{l_2}{2}\right)
\end{aligned}$$

$$= hf_2(0.1, 0.99, -0.101)$$
$$= 0.2(xz - y^2)$$
$$= 0.2[0.1 \times (-0.101) - (0.99)^2]$$
$$= 0.2[-0.0101 - 0.9801] = -0.2 \times 0.9902 = -0.198$$
$$k_4 = hf_1(x_0 + h, y_0 + k_3, z_0 + l_3)$$
$$= (hf_1(0 + 0.2, 1 - 0.02, 0 - 0.198))$$
$$= .2 \times (-0.198) = -0.0396$$
$$l_4 = hf_2(0.2, 0.98, -0.198)$$
$$= 0.2(xz - y^2)$$
$$= 0.2[0.2 \times -0.190 - 0.98^2]$$
$$= 0.2(-0.0396 - 0.9604) = -0.2$$
$$k = \frac{1}{6}[k_1 + 2(k_2 + k_3) + k_4]$$
$$= \frac{1}{6}[0 + 2(-0.02 - 0.0202) - 0.040] = -0.02$$
$$l = \frac{1}{6}[l_1 + 2(l_2 + l_3) + l_4]$$
$$= \frac{1}{6}[-0.2 + 2[-0.202 - 0.198] - 0.2] = -0.2$$
$$\therefore \quad y = y_0 + k \,;\, z = z_0 + l$$
$$= 1 - 0.02 \,;\, z = 0 - 0.2$$
$$= 0.98 \,;\, z = -0.2$$
$$\left.\begin{aligned} z &= 0.2 \\ y &= 0.98 \\ z &= -0.2 \end{aligned}\right\} \textbf{ Ans.}$$

5.21 ERRORS, PRECISION AND ACCURACY

All measurements are subject to the following errors:

1. Systematic errors: This type of errors affect all measurements alike. Such errors are caused mainly due to imperfections in the construction or adjustment of instruments and observer's attitude. The systematic errors are usually determinate and can be removed by applying the proper corrections.

2. Mistakes or Blunders: Mistakes or blunders are the large errors due to negligence of the observers or carelessness of the observers while reading of the measuring instruments/or by faulting recording of the readings. Mistakes are generally committed by reading to the wrong scale, reading vernier backward, making a miscount in observations which may involve recording the reading in the wrong column. Mistakes do not follow any law and can be avoided by only constant vigilance and careful checking by the observer.

3. Accidental errors: Such errors are unknown and indeterminate but they are generally small and follow the law of chances.

Precision of measurement: Measurements obtained by instruments give directly the quantity desired. Direct measurements are usually made by reading a scale graduated in terms of desired unit.

In direct measurements are those in which the quantity measured are not given directly by observations or readings taken. Indirect measurements are calculated from the observed readings. Thus, in an indirect measurement, the desired quantity is a function of one or more directly measured quantities.

For example if want to calculate the area of a triangular plot of land, we only measure two sides b and c and also the included angle A.

$$\text{Area} = \frac{1}{2}bc \sin A$$

To find the length of side a, we use cosine formula is

$$\text{Side } a = \sqrt{b^2 + c^2 - 2bc \cos A}$$

Precision and Accuracy: The two words 'precision' and 'accuracy' differ in meanings when they are used in measurements. Precision refers to the accidental errors whereas a precise measurement would·be free from accidental errors. On the other hand an accurate measurement would be free from all kinds of errors. Ignoring the mistakes, the systematic error is the difference between the precise value and an accurate or true value of the quantity measured by using more precise instruments and method whereas the precision can be more and only by using more care while making measurements by the observer.

❑❑❑

6 Linear Programming

6.1 INTRODUCTION

Linear programming or linear optimization consists of the method for solving optimization problems in which object function f is a linear function of the control variables $x_1, x_2, \ldots x_n$ where the domains of these variables are defined by the constraints. The linear programming technique is widely used to tackle a number of industrial, economic, marketing and distribution problems. More over, the technique of linear programming is widely applied to important areas of product mix, industries, steel industries and food processing industries. In fact, the linear programming technique may be applied to any situation, where a linear function of variables is required to be optimized subject to a set of linear equations or inequalities.

The problems which seek to maximise/minimise profit/cost form a general class of problem called *optimisation problems*. The very important class of optimization is the linear programming problem (L.P.P.) Linear programming problems have gained great importance because of their wide applicability in the present day industry, commane, managements, etc.

We shall discuss the solution of linear programming problems by the graphical method which provides a great deal of insight in the basic concepts.

6.2 FORMULATION OF LINEAR PROGRAMMING PROBLEMS

The problem is first presented in a linear programming form which requires defining the variables involved in the problem and also establishing relationships between them and further formulating the object function and the constraint. The technique of linear programming is explained with the help of a solved example to familiarise the students.

Example 6.1. *A manufacturer manufactures and sells two types of lamps L_1, L_2, the profit being ₹15 and ₹10 respectively. The manufacturing process involves two workers W_1 and W_2 who are available for this type of work 100 hours and 80 hours permonth, respectively. W_1 assembles L_1 in 20 minutes and L_2 in 30 minutes. W_1 paints L_1 in 20 W_2 paints and L_2 in 10 minutes. Assuming that all the lamps are sold in the market, determine the production of the lamps for obtaining the maximum profit.*

Solution: Let L_1 and L_2 represent the number of two types of lamps. The given general linear optimization problem is converted into normal form:

$$15\, L_1 + 10\, L_2 = Z, \text{ the maximum}$$

Constraints:

$$\frac{1}{3} L_1 + \frac{1}{2} L_2 \geq 100 \quad \text{...(1)}$$

$$\frac{1}{3} L_1 + \frac{1}{6} L_2 \geq 80 \quad \text{...(2)}$$

For non-negative values of L_1 and L_2

$$L_1 \geq 0 \ ; \ L_2 \geq 0$$

From eqn. (1), we get

If $L_1 = 0,\quad L_2 = 200$

If $L_2 = 0,\quad L_1 = 300$

From eqn. (2), we get

If $L_1 = 0,\quad L_2 = 480$

If $L_2 = 0,\quad L_1 = 240$

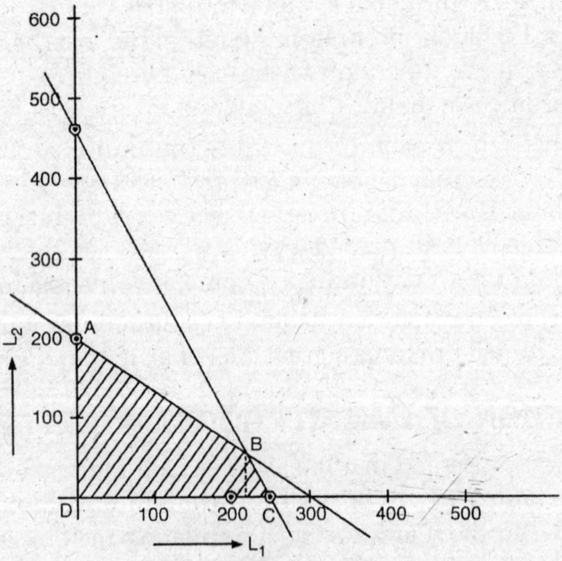

Fig. 6.1

Plot the graphs of two straight lines $\frac{1}{3} L_1 + \frac{1}{6} L_2 = 100$ and $\frac{L_1}{3} + \frac{L_2}{6} = 80$ (Fig. 6.1)

The coordinates of vertices of feasible area are:

	L_1	L_2
A	0	200
B	210	60
C	240	0
D	0	0

Calculation of maximum profits:

A : $15 \times 0 + 10 \times 200 = ₹\,2000$

B : $15 \times 210 + 10 \times 60 = ₹\,3750\ f_{max}\ (210, 60)$ **Ans.**

A : $15 \times 240 + 10 \times 0 = ₹\,3600$

D : $15 \times 0 + 10 \times 0 = ₹\,0$

Example 6.2. *Solve the following linear programming problem graphically:*

$$\text{Maximise} = 6x + 5y$$

Subject to:
$$5x + 2y \le 10$$
$$3x + 5y = 15$$
$$x, y \ge 0$$

Solution: Given constraints:

$$3x + 5y \le 15;$$
$$5x + 2y \le 10$$

For the inequality $3x + 5y \le 15$, the corresponding linear equation is $3x + 5y = 15$.

When $y = 0, x = 5$, when $x = 0, y = 3$.

For the inequality $5x + 2y \le 10$, the corresponding linear equation is $5x + 2y = 10$

When $y = 0, x = 2$; when $x = 0, y = 5$.

The graph of these inequalities is shown in Fig. 6.2.

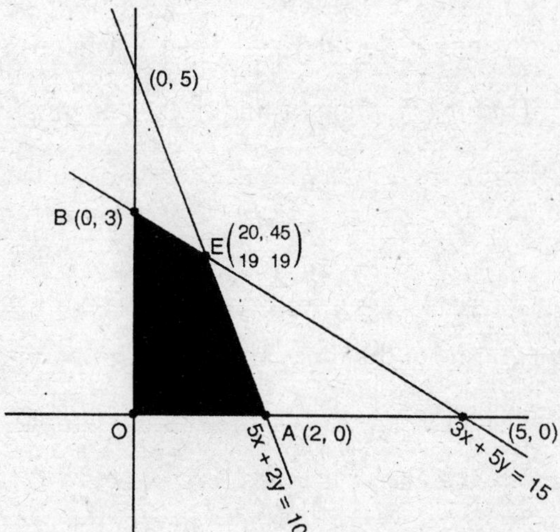

Fig. 6.2

Calculation of the coordinates of the point of intersection of equations:

$$3x + 5y = 15 \quad \text{...}(i)$$
$$5x + 2y = 10 \quad \text{...}(ii)$$

Multiplying equation (*i*) by 5 and equation (*ii*) by 3 we get

$$15x + 25y = 75 \quad \text{...}(iii)$$
$$15x + 6y = 30 \quad \text{...}(iv)$$

$$\therefore \quad 19y = 45 \quad \text{or} \quad y = \frac{45}{19}$$

Multiplying equation (*i*) by 2 and equation (*ii*) by 5 we get

$$6x + 10y = 30$$
$$25x + 10y = 50$$

$$\therefore \quad 19x = 20 \quad \text{or} \quad x = \frac{20}{19}$$

The cornor points of fasible region are:

O	(0, 0);
A	(2, 0);
E	$\left(\frac{20}{19}, \frac{45}{19}\right)$
B	(0, 3)

The values of the objective function at these corner points are:

Corner point	Value of the objective function $z = 6x + 5y$
O	$6 \times 0 + 5 \times 0 = 0$
A	$6 \times 2 + 5 \times 0 = 12$
E	$6 \times \frac{20}{19} + 5 \times \frac{45}{19} = \frac{345}{19} > 18$
B	$6 \times 0 + 5 \times 3 = 15$

$\therefore$ Maximum value of Z is at E$\left(\frac{20}{19}, \frac{45}{19}\right)$ **Ans.**

Example 6.3. *Find the minimum value of 3x + 5y graphically subject to the constraints:*

$$-2x + y \leq 4$$
$$x + y \geq 3$$
$$x - 2y \leq 2$$
$$x \geq 0, y \geq 0$$

Solution: The graphs of the inequalities are shown in Fig. 6.3, the corner points of the shaded region are: A$\left(\frac{8}{3}, \frac{1}{3}\right)$, B(0, 3), C(0, 4).

The value of $Z = 3x + 5y$

at $$A = 3 \times \frac{8}{3} + 5 \times \frac{1}{3} = \frac{20}{3}$$

at B = $3 \times 0 + 5 \times 3 = 15$

at C = $3 \times 0 + 5 \times 4 = 20$

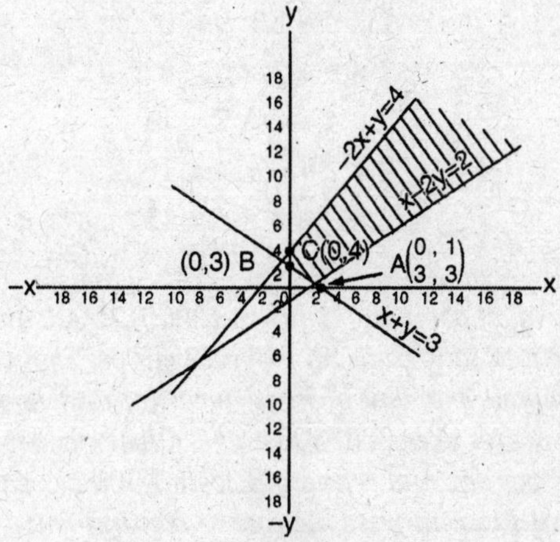

Fig. 6.3

since, the minimum value of these three values is 20/3

i.e., at A, $\left(x = \frac{8}{3};\ \ y = \frac{1}{3}\right)$

Hence, minimum value of Z = 20/3 **Ans.**

Example 6.4. *Maximize P = 2x + 3y subject to the following constraints:*

$$2x + 2y \leq 10$$
$$x + y \leq 6$$
$$x \leq 4$$

and $$x \geq 0, y \geq 0$$

Solution: The graphs of the inequalities are in Fig. 6.4.

OABCD is the feasible region

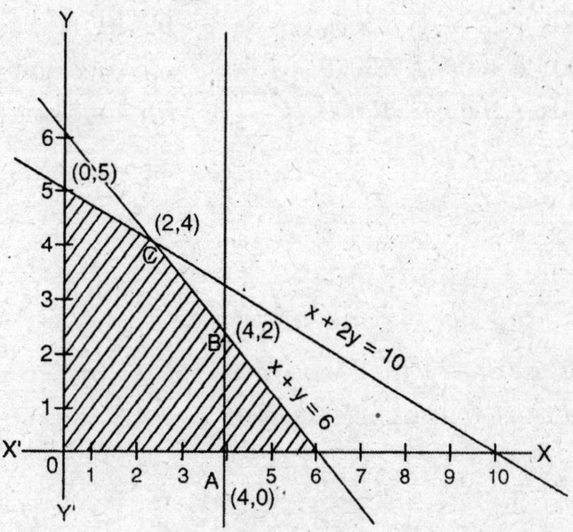

Fig. 6.4

Corner points showing the shaded region are:

0(0, 0), A(4, 0), B(4, 2), C(2, 4), D(0, 5)

$$P_0 = 2 \times 0 + 3 \times 0 = 0$$

$$P_A = 2 \times 4 + 3 \times 0 = 8$$

$$P_B = 2 \times 4 + 3 \times 2 = 14$$

$$P_C = 2 \times 2 + 3 \times 4 = 16 \quad \text{...(Maximum value)}$$

$$P_D = 2 \times 0 + 3 \times 5 = 15$$

$\therefore$ Maximum value of P = 16 when $x = 2$, $y = 4$. **Ans.**

Example 6.5. *An oil company requires 13000, 20000 and 15000 barrels of high grade, medium grade and low grade oil respectively per month. Refinery A produces 100, 300 and 200 barrels per day of high, medium and low grade oil respectively whereas the refinery B produces 200, 400 and 100 barrels per day respectively. If the refineryA costs ₹400 per day and refinery B costs ₹300 per day to operate, how many days should each be run to minimise the cost of requirement?*

Solution:

Refinery	High grade	Medium grade	Low grade	cost/day
A	100	300	200	₹ 400
B	200	400	100	₹ 300

Requirement of barrels 13000 high grade, 20,000 medium grade 15000 low grade

Let x be the number of days the refinery A runs

y be the number of days the refinery B runs

The linear programming problem now states the following:

Minimise $z = 400x + 300y$

Subject to the constraints:

$$100x + 200y \geq 13000 \quad \text{for high grade}$$

$$300x + 400y \geq 20000 \quad \text{for medium grade}$$

$$200x + 100y \geq 15000 \quad \text{for low grade}$$

$$x, y \geq 0$$

The inequalities reduce to:

$$x + 2y \geq 130 \quad \text{...}(i)$$

$$3x + 4y \geq 200 \quad \text{...}(ii)$$

$$2x + y \geq 150 \quad \text{...}(iii)$$

Plotting the graphs of the inequalities (Fig. 6.5), we get

Corner points of the feasible region

A[0, 150], $B\left[\frac{170}{3}, \frac{110}{3}\right]$, C(130, 0)

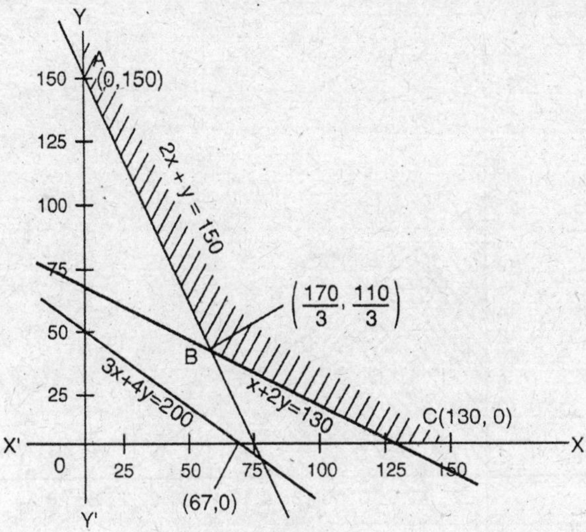

Fig. 6.5

$$\therefore \quad Z_A = 400 \times 0 + 300 \times 150 = 45{,}000$$

$$Z_B = 400 \times \frac{170}{3} + 300 \times \frac{110}{3} = 33{,}667$$

$$Z_C = 400 \times 130 + 300 \times 0 = 52{,}000$$

$\therefore$ Minimum cost is at $x = \frac{170}{3}, y = \frac{110}{3}$ **Ans.**

Example 6.6. *A firm makes items A and B and the total number of items it can make in a day is 24. It takes one hour to make an item of A and only half an hour to make an item of B. The maximum time available per day is 16 hours. The profit on an item of A is ₹300 and on one item of B is ₹160. How many items of each type should be produced to maximize the profit? Solve the problem graphically.*

Solution: (Fig. 3)

Let number of items of type A $= x$

Number of items of type B $= y$

Linear programming problem states:

Maximize $\quad Z = 300x + 160y$

Subject to constraints:

$$x + y \leq 24$$
$$2x + y \leq 32$$
$$x, y > 0$$

Graph of these inequalities is shown in Fig. 6.6.

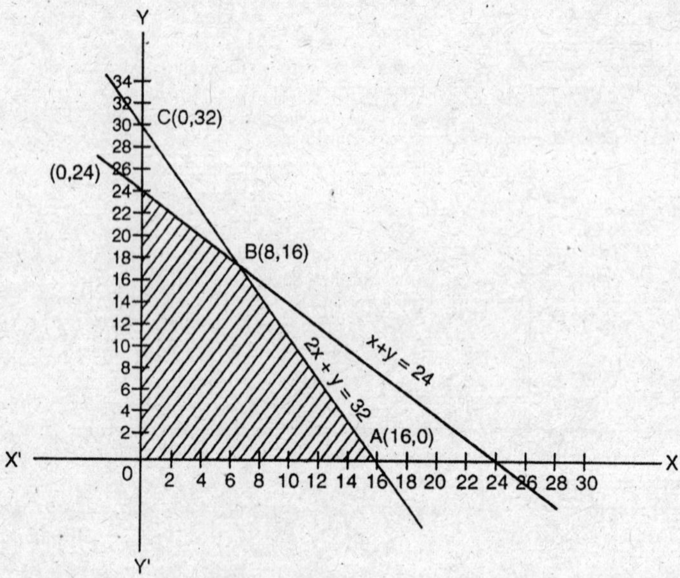

Fig. 6.6

Corners of the feasible region are:

A(16, 0); B(8, 16); C(0, 32)

$$Z_A = 300 \times 16 + 160 \times 0 = 4800$$
$$Z_B = 300 \times 8 + 160 \times 16 = 4960 \quad \text{Maximum}$$
$$Z_C = 300 \times 0 + 160 \times 24 = 3840$$

∴ Maximum profit is when $x = 9, y = 16$ **Ans.**

Example 6.7. *A company manufacturers two types of toys A and B. Type A requires 5 minutes each for cutting and 10 minutes each for assembling. Type B requires 8 minutes each for cutting and 8 minutes each for assembling. There are 3 hours available for cutting and 4 hours available for assembling in a day. The profit is ₹50 each on type A and ₹60 each on type B. How many toys of each type should the company manufacture in a day to maximise the profit?*

Solution: Let x and y be the number of toys of types A and B respectively

To maximise $z = 50x + 60y$ subject to the following constraints;

$$5x + 8y \leq 3 \times 60$$
$$10x + 8y \leq 4 \times 60$$
$$x \geq 0; y \geq 0$$

Now draw the graphs of the following equations:

$$5x + 8y = 180$$
$$10x + 8y = 240$$

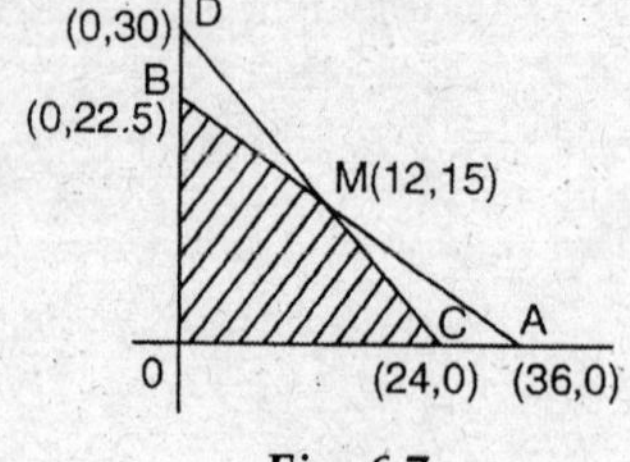

Fig. 6.7

The point of intersection M of the lines AB and CD is M(12, 15)

The feasible region OCMB (Fig. 6.7) is shaded

Value of $z = 50x + 60y$

at C(24, 0) $\quad z = 50 \times 24 + 60 \times 0 = 1200$

at M(12, 15) $\quad z = 50 \times 12 + 60 \times 15 = 1500$

at B(0, 22.5) $\quad z = 50 \times 0 + 60 \times 22.5 = 1350$

For obtaining the maximum profit, the company should manufacture 12 toys of type A and 15 toys of type B. **Ans.**

Example 6.8. *A diet for a sick person must contain at least 4000 units of vitamins, 50 units of minerals and 1400 calories. Two foods X and Y are available at a cost of ₹4 and ₹3 per unit respectively. One unit of the food X contains 200 units of vitamins, 1 unit of minerals and 40 calories, whereas one unit of food Y contains 100 units of vitamins, 2 units of minerals and 40 calories. Find what combination of X and Y should be used to have least cost, satisfying the requirements.*

Solution:

	Vitamins	*Minerals*	*Calories*	*Cost/Kg.*
X	200	1	40	₹ 4
Y	100	2	40	₹ 3

Let the mixture contain x kg. of food X and y kg. of food Y.

Optimization function $z = 4x + 3y$

subject to constraints:

$$200x + 100y \geq 4000 \Rightarrow 2x + y \geq 40 \quad \text{...(i)}$$

$$x + 2y \geq 50 \Rightarrow x + 2y \geq 50 \quad \text{...(ii)}$$

$$40x + 40y \geq 1400 \Rightarrow x + y \geq 35 \quad \text{...(iii)}$$

$$x \geq 0,\ y \geq 0$$

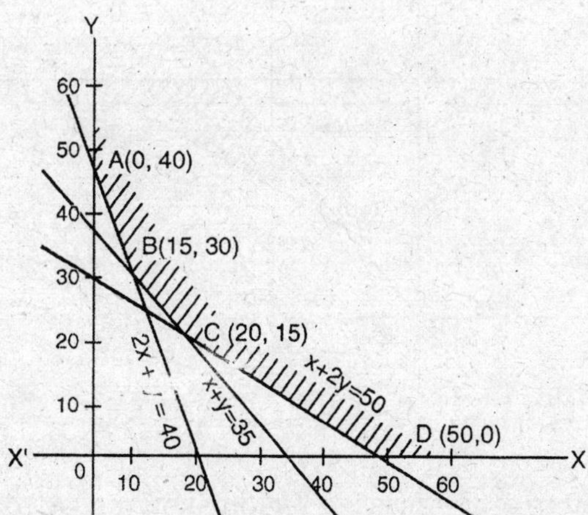

Fig. 6.8

Graph of the inequalities is shown in Fig 6.8.

Coordinates of cardinal points:

Point A	(0, 40)
Point B	(15, 30)
Point C	(20, 15)
Point D	(50, 0)

Putting the coordinates of the cardinal points, we get

$$Z(0, 40) = 4 \times 0 + 3 \times 40 = 1200$$
$$Z(5, 30) = 4 \times 5 + 3 \times 30 = 110$$
$$Z(20, 15) = 4 \times 20 + 3 \times 15 = 125$$
$$Z(50, 0) = 4 \times 50 + 3 \times 0 = 200$$

$\therefore$ Z is minimum at (5, 30)

$\therefore$ 5 units of food X and 30 units of food Y be mixed to get the needed nutrients. at the least cost. **Ans.**

Example 6.9. *Solve the following linear programming problem graphically. Minimise* $Z = x - 5y + 20$ *subject to constraints:*

$$x - y \geq 0$$
$$-x + 2y \geq 2$$
$$x \geq 3$$
$$y \geq 4$$
$$x, y \geq 0$$

Solution: (Fig. 5)

Draw the graphs of the given inequalities as shown in Fig. 6.9.

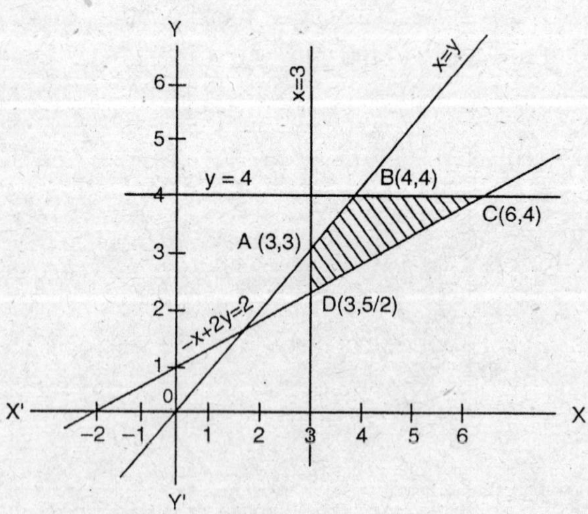

Fig. 6.9

Cardinal points of the feasible region shown shaded are:

A(3, 3) $\qquad Z = x - 5y + 20 = 3 - 15 + 20 = 8$

B(4, 4) $Z = x - 5y + 20 = 4 - 20 + 20 = 4$...Minimum

C(6, 4) $Z = x - 5y + 20 = 6 - 20 + 20 = 6$

D(3, 5/2) $Z = x - 5y + 20 = 3 - 12\frac{1}{2} + 20 = 10.5$

$\therefore$ Z is minimum at (4, 4) **Ans.**

Example 6.10. *A retired person has ₹70,000 to invest and two types of bonds are available in the market for investment. First type of bond yields an annual income of 8% on the amount invested and the second type of bond yields 10% per annum. As per norms he has to invest minimum of ₹10,000 in the first type and not more than ₹30000 in the second type. How should he plan his investment, so as to get maximum return, after one year of investment?*

Solution: Let the investment in the first type of bonds = ₹ x;

the investment in the second type of bonds = ₹ y.

$\therefore$ The linear programming problem becomes:

Maximise $Z = \frac{8x}{100} + \frac{10y}{100}$

Subject to constraints:

$$x + y \leq 70{,}000 \quad \text{...(i)}$$
$$x \geq 10{,}000 \quad \text{...(ii)}$$
$$y \leq 30{,}000 \quad \text{...(iii)}$$
$$x \geq 0, y \geq 0 \quad \text{...(iv)}$$

Plot the graphs of inequalities (*i*) to (*iv*) as shown in Fig. 6.10.

When $y = 0$, $x = 70{,}000$; When $x = 0$, $y = 70{,}000$

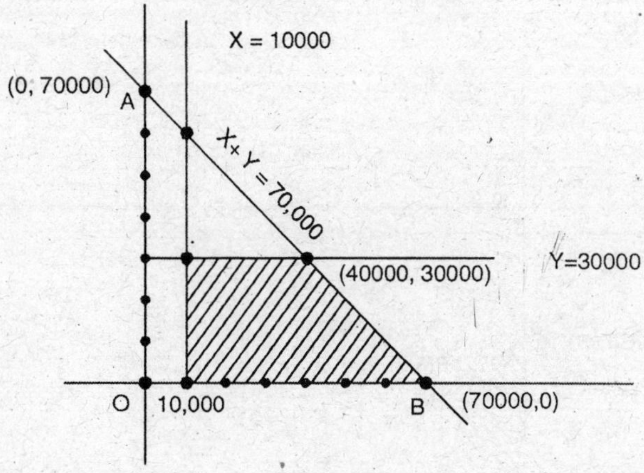

Fig. 6.10

Value of Z at cardinal points is:

$$A = \frac{8 \times 10000}{100} + \frac{10 \times 30000}{100} = 800 + 3000 = ₹\ 3{,}800$$

$$B = \frac{8 \times 40000}{100} + \frac{10 \times 30000}{100} = 3200 + 3000 = ₹\ 6{,}200$$

$$C = \frac{8 \times 70000}{100} + \frac{10 \times 0}{100} = 5600 + 0 = ₹\ 5{,}600$$

$$D = \frac{8 \times 10000}{100} + \frac{10 \times 0}{100} = 800 + 0 = ₹\ 800$$

The maximum value of Z is at B

∴ For maximum returns he should invest

₹ 40,000 in the first type of bonds

and ₹ 30,000 in the second type of bonds. **Ans.**

Example 6.11. *A producer has 30 and 17 units of labour and capital respectively, which he can use to produce two types of goods X and Y. To produce one unit of X, 2 units of labour and 3 units of capital are required. Similarly, 3 units of labour and 1 unit of capital is required to produce one unit of Y. If X and Y are priced at ₹100 and ₹120 per unit respectively, how should the producer use his resources to maximize the total revenue? Solve the problem graphically.*

Solution: Let x be the number of units of X type goods

y be the number of units of Y type goods

The given information may be tabulated as under:

	X	*Y*	*Available units*
Labour	2	3	30
Capital	3	1	17
Revenue unit	Rs. 100	Rs. 120	

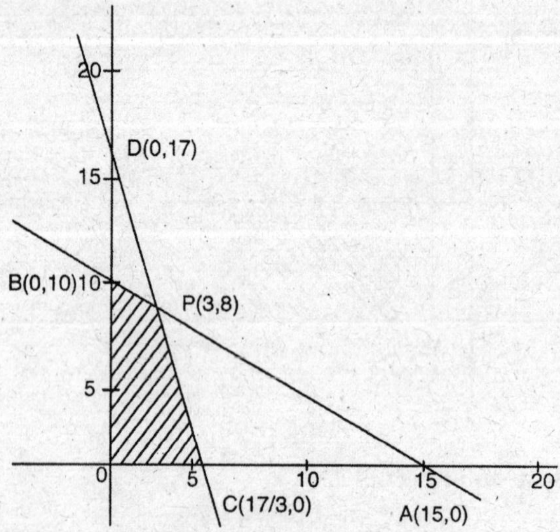

Fig. 6.11

Keeping in view, the revenue we shall maximise: Z = 100X + 120Y subject to constraints:

$$2x + 3y = \leq 30$$
$$3x + y = \leq 17$$
$$x \geq 0; y \geq 0$$

Plotting the graphs of the inequalities as shown in Fig. 6.11, we have The coordinates of the intersection points are as under:

Extreme point	*Coordinates*	*Value of Z= 100x + 120y*
0	(0, 0)	0
C	(17/3, 0)	1700/3
P	(3, 8)	1260
B	(0, 10)	1200

The value of Z is maximum at P(3, 8), having a value of ₹ 1260 **Ans.**

Example 6.12. *A firm deals with two kinds of fruit juices - pine apple and orange juice. These are mixed and two types of mixtures are obtained, which are sold as soft drinks A and B. One tin of A needs 4 litres of pine apple juice and 1 litre of orange juice. One tin of B needs 2 litres of pine apple and 3 litres of orange juice. The firm has only 46 litres of pineapple juice and 24 litres of orange juice. Each tin of A and B is sold at a profit of ₹4/- and ₹3/- respectively. How many tins of juice A and B should the firm produce to maximise profit? Formulate the linear programming problem and solve it graphically using iso profit line approach.*

Solution: Given: A tin contains 4 litres of pineapple and 1 litre of orange juice

B tin contains 2 litres of pineapple and 2 litres of orange juice.

Let the number of tins soft drink A = x, soft drink B = y.

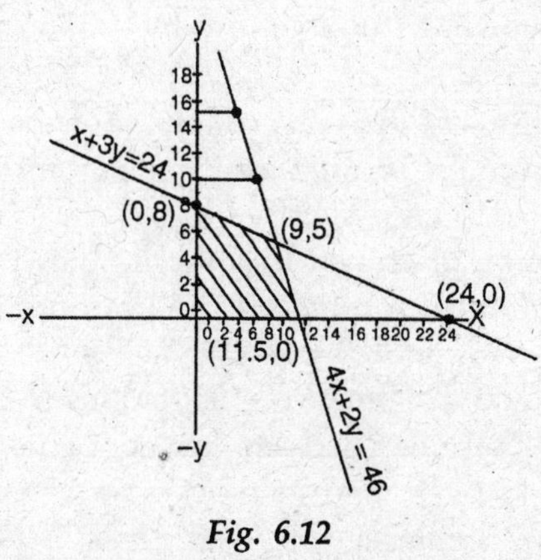

Fig. 6.12

The required linear programming problems is:

To maximise the profit P = $4x + 3y$

subject to: $4x + 2y \leq 46$

$x + 3y \leq 24$

$x \geq 0, y \geq 0$

One inequality orOther inequality or

Iso-profit line is $4x + 2y = 46$ Iso-profit line is $x + 3y = 2$

Plot the graphs as shown in Fig. 6.12.

$y = 0$	$x = 11.5$	A = (11.5, 0)	Put	$y = 0,$	$x = 24$
$y = 5$	$x = 9$	B = (9.5)	$y = 5,$	$x = 9$	
$y = 10$	$x = 6.5$	C = (0, 8)	$y = 10$	$x = -6$	
$y = 15$	$x = 4$				

Using iso-profit lines, we get $x = 9$, $y = 5$ for maximise profit. **Ans.**

Example 6.13. *A firm manufactures two types of products A and B and sells them at a profit of ₹5 per unit of type A and ₹3 per unit of type B. Each product is processed on two machines M_1 and M_2. One unit of type A requires one minute of processing time on M_1 and two minutes of processing time on M_2, where as one unit of type B requires one minute of processing time on M_1 and one minute on M_2. Machines M_1 and M_2 are respectively available for atmost 5 hours and 6 hours in a day. Find out how many units of each type of product should the firm produce a day in order to maximise the profit. Solve the problem graphically.*

Solution: The given data may be tabulated as under:

Product / *Machines*	*A*	*B*	*Time available (minutes)*
M_1	1	1	300
M_2	2	1	360
Profit	5	3	

Let the firm manufacture x units of product A and y units of product B to maximise its profit.

$\therefore$ Total profit = ₹ $(5x + 3y)$ or total revenue $Z = 5x + 3y$

Total number of minutes required for x units of product A and y units of product B machine $M_1 = x \times 1 + y \times 1 = x + y$

But, available time on machine M_1 = 300 minutes $\therefore x + y \leq 300$

Total number of minutes required for x units of product A and y units of product of B on machine $M_2 = x \times 2 + y \times 1 = 2x + y$

But, available time on machine M_2 = 360 minutes $\therefore 2x + y \leq 360$

As the number of both the products can not be negative,

$\therefore$ $x \geq 0$ and $y \geq 0$

The linear problem becomes:

$$Z = 5x + 3y$$

subject to the constraints:

$$x + y \leq 300$$
$$2x + y \leq 360$$
$$x \geq 0 \text{ and } y \geq 0$$

We now draw iso profit lines for $x + y = 300$ and $2x + y = 360$

The shaded portion shall satisfy the inequalities:

The Corner point method: The coordinates of the vertices O, C, E and A are O(0, 0), C(0, 300), E(60, 240), A(180, 0)

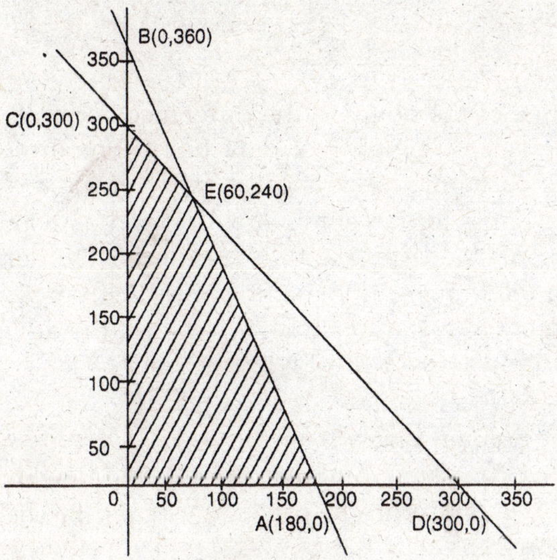

Fig. 6.13

The coordinates of E are obtained by solving $x + y = 300$ and $2x + y = 360$

I(0, 0)	5(0) + 3(0)	= 0
C(0, 300)	5(0) + 3(300)	= 900
E(60, 240)	6(60) + 3(240)	= 1020 Maximum value.
A(180, 0)	5(180) + 3(0)	= 900

From the above we note that Z is maximum when $x = 60$, $y = 240$.

The total profit is maximum when the firm manufactures 60 units of product A and 240 units of product B. **Ans.**

EXERCISE

1. (*a*) Applying Runge-Kutt fourth order formula find the solution of the equation

$$\frac{dy}{dx} = -(1-x)^2y^2$$ for one step only with initial values $x_0 = 0$, $y(x_0) = 2$, and the interval, $h = 0.2$.

(*b*) Write a computer program for Runge-Kutta fourth-order method for $x_0 = 0$ to $x_n = 2$ for the above equation.

2. (*a*) What is the transpose of a matrix? Write a program which will read an M × N matrix and print out the transpose matrix.
   (*b*) Explain the modified Euler's method of numerical integration. Write a subroutine which can do numerical integration by Euler's and modified Euler's methods.
3. (*a*) Write a program to read 10 × 10 matrix. Assume that all the elements of a row are entered in the same line. Print the maximum element of the matrix.
   (*b*) The harmonic mean of a set N deta points $x_1, x_2 \ldots x_8$ is defined as:

$$H_{MEN} = N/\sum_{i=1}^{N} \frac{1}{x_1}$$

   The range of the points is the difference between smallest member and the largest member of data point. Prepare a sub-program to find the harmonic mean and the range of the array.
4. Explain the Gauss Jordan elimination method of evaluating the inverse of a matrix. Write an application program for the purpose.
5. (*a*) Explain the Runge-Kutta method and Taylor method for integration of differential equations briefly. What is the advantage or Runge-Kutta formulas over the Taylor method?
   (*b*) Apply the local Taylor ....
6. (*a*) Define a column matrix.
   (*b*) Draw the flow char program in FORTRAN which will be (M, N) and E(N, L) and print the product of A and B. The order of A and B will not exceed 20 × 20.
7. (*a*) Describe the trapezoidal rule for numerical integration of a continuous function. Write a computer program for this method.
   (*b*) Describe the Gaussian elimination method of solving simultaneous algebraic equations. Take any three sets of simultaneous equations with three unknowns and solve by this method.
8. (*a*) What are the different MOVE statements in COBOL? What are their functions?
   (*b*) Are ROUNDED and ON SIZE ERROR options permitted in COBOL? What do they perform?
   (*c*) What are the hierarchy rules for conditional statements in COBOL?
   (*d*) How is a COBOL program terminated? What are the two statements that must be coded before the program is complete?
9. Solve the following simultaneous equations by matrix method:

$$\begin{aligned} x - 2y + 3z &= 2 \\ 2x + 0 - 3z &= 3 \\ x + y + z &= 0 \end{aligned}$$

10. Solve the following simultaneous equations by matrix method:

(*i*)
$$\begin{aligned} x + xy + z &= 6 \\ 2x + 3y + 3z &= 17 \\ x - y + z &= 2 \end{aligned}$$

(*ii*)
$$\begin{aligned} 4x - y + z &= 4 \\ x + 2y + z &= 4 \\ 3x - 2y + 3z &= 4 \end{aligned}$$

11. Determine the largest 'eigon value' of the matrix

$$A = \begin{bmatrix} 10 & -2 & 1 \\ -2 & 10 & -2 \\ 1 & -2 & 10 \end{bmatrix}$$

12. (*a*) What are the sources of error in numerical computations? Explain what you understand by the terom 'Convergence'.

(*b*) Solve the equation $\frac{dy}{dx} = 2x \times y$

With the initial conditions $x_0 = 1$m $y_0 = 1$ using Runge-Kutta method. Take the internal between successive values of x as 0.1.

13. Find the inverse of the following matrices

(*a*) $\begin{bmatrix} 3 & 5 & 7 \\ 1 & -2 & 1 \\ 3 & 2 & 4 \end{bmatrix}$ (*b*) $\begin{bmatrix} 1 & 5 & 3 \\ 0 & 2 & 5 \\ 3 & 2 & 1 \end{bmatrix}$ (*c*) $\begin{bmatrix} 1 & 6 & 4 \\ 0 & 2 & 3 \\ 0 & 1 & 2 \end{bmatrix}$ (*d*) $\begin{bmatrix} 3 & 2 & 3 \\ 1 & 2 & 3 \\ 3 & 2 & 1 \end{bmatrix}$

14. Solve the following simultaneous equations by Gauss-Elimination method.

$$A = \begin{bmatrix} 5 & -2 & 1 \\ 7 & 1 & -5 \\ 3 & 7 & 4 \end{bmatrix}$$

15. Solve the following simultaneous equations by Gauss siedel-iterative method

$$9x_1 - 3x_2 - 4x_3 = 2$$
$$2x_1 + x_2 + 3x_3 = 6$$
$$2x_1 - 3x_2 + 6x_3 = 5$$

16. Show that the matrix

$$A = \begin{bmatrix} 3 & 5 & 4 \\ 4 & -3 & 2 \\ 2 & 1 & 3 \end{bmatrix}$$

satisfies its characteristics equation and hence. Determine A^{-1}.

17. Bring out the meaning of the following terms used in memorical analysis:
   (*i*) Forward difference; (*ii*) Divided difference;
   (*iii*) Truncation error; (*iv*) Difference equations.

18. (*a*) Define the following errors:
   (*i*) Relative error; (*ii*) Absolute error; (*iii*) Percentage error.

   (*b*) Describe the method of solution of a differential equation with initial values, by Enter's method.

19. Define precisely and explain with the help of simple examples the following error terms used in numerical analysis:
   (*i*) Absolute error (*ii*) Relative error
   (*iii*) Round of error (*iv*) Propagation error
   (*v*) In herent error

□□□